ROUTLEDGE LIBRARY EDITIONS:
HUMAN GEOGRAPHY

Volume 12

PLURALISM AND
POLITICAL GEOGRAPHY

PLURALISM AND POLITICAL GEOGRAPHY

People, Territory and State

Edited by
NURIT KLIOT AND STANLEY WATERMAN

LONDON AND NEW YORK

First published in 1983 by Croom Helm Ltd

This edition first published in 2016
by Routledge
2 Park Square, Milton Park, Abingdon, Oxon OX14 4RN

and by Routledge
711 Third Avenue, New York, NY 10017

Routledge is an imprint of the Taylor & Francis Group, an informa business

British Library Cataloguing in Publication Data
A catalogue record for this book is available from the British Library

ISBN: 978-1-138-95340-6 (Set)
ISBN: 978-1-315-65887-2 (Set) (ebk)
ISBN: 978-1-138-95896-8 (Volume 12) (hbk)
ISBN: 978-1-315-66092-9 (Volume 12) (ebk)

Publisher's Note
The publisher has gone to great lengths to ensure the quality of this reprint but
points out that some imperfections in the original copies may be apparent.

Disclaimer
The publisher has made every effort to trace copyright holders and would welcome
correspondence from those they have been unable to trace.

PLURALISM AND POLITICAL GEOGRAPHY

PEOPLE, TERRITORY AND STATE

Edited by
NURIT KLIOT and STANLEY WATERMAN

CROOM HELM
London & Canberra
ST. MARTIN'S PRESS
New York

© 1983 N. Kliot and S. Waterman
Croom Helm Ltd, Provident House, Burrell Row,
Beckenham, Kent BR3 1AT

British Library Cataloguing in Publication Data

Pluralism and political geography.
    1. Nationalities, Principle of   2. Anthropo-
geography
    I. Kliot, Nurit   II. Waterman, Stanley
    323.6   GF50
    ISBN 0–7099–1537–3

Library of Congress Cataloging in Publication Data
Main entry under title:

Pluralism and political geography.

    Includes bibliographies and index.
    1. Geography, political––Addresses, essays, lectures.
2. Pluralism (social sciences)––Addresses, essays,
lectures. I. Kliot, Nurit. II. Waterman, Stanley.
JC319.P58   1983   320.1'2   83–6212
ISBN 0–312–61766–6

Printed and bound in Great Britain by
Biddles Ltd, Guildford and King's Lynn

CONTENTS

CONCLUSION

Index

For our families

PREFACE

In January 1982 a week-long seminar on Contemporary Trends in Political Geography was held in the Department of Geography at the University of Haifa. For eight days, some 30 geographers from seven countries discussed political geography in formal sessions, on field-trips to Jerusalem and Israel's northern borders, and over breakfast, lunch and dinner.

This book had its origins and inspiration in that seminar. It does not represent the proceedings of the seminar - only a portion of the papers presented then are presented here; several of the essays that appear in the book were invited following the seminar and several others originally given at the Haifa seminar have been reworked and rewritten so as to better fit the concept of pluralism which is the common theme of the book.

The preparation of the book was affected by the events in Lebanon during the second half of 1982. Not only was the editing process itself actually delayed by the war and the active service of one of the editors, but a closer look at Lebanon threw additional light on the problems of what is one of the most plural societies in the world, where groups which perceive their own uniqueness, individuality and personality are forced by history, geography and other circumstances to live side by side, and where the political structure of the country is probably more institutionalised in relation to pluralism than in any other state. The hazards of the plural society and plural state beyond the northern frontier of Israel proved to be an educational, but somewhat sad, eye-opener to a pair of innocent academics.

No book is prepared without help from many people. To begin to thank everyone who proffered advice and aid is difficult; to do so is probably hazardous, as the probability of omissions is high.

Our first thanks are to our friends and colleagues in the Department of Geography at the University of Haifa and to the University's institutions who provided us with the office space, word-processor, secretarial and cartographic facilities with which to work. The Reuben and Edith Hecht Foundation, The Institute of Middle Eastern Studies and the Faculty of Social

Sciences at the University of Haifa, the Israel National Academy
of Sciences and the National Research Council provided us with
the bulk of financial support for the initial seminar and thus
for the preparation of the book. Aliza Gold and Aviva Shalit
provided prompt and accurate drafting service; the Photography
Laboratory at the Technion was responsible for the excellent
photographic work on the maps. Saul Cohen, Stanley Brunn, Neville
Douglas and Peter Taylor provided us with support and encouragement
in the early stages of preparation but, of course, they bear
none of the responsibility for the editorial decisions that were
eventually taken and for what appears between the covers of the
book.

N.K. & S.W.
Haifa

Chapter 1
INTRODUCTION
Nurit Kliot and Stanley Waterman

This volume explores pluralism as a research topic in contemporary political geography. As the science studying 'areal differentiation' or 'spatial variation', geographers have for long understood the relevance and importance of the theme of pluralism. As such, this book not only presents recent work on people, territory and state from pluralist perspectives but also shows the great variation that exists among the community of researchers who study pluralism.

INTRODUCTION

The most common definitions of pluralism focus attention on two major meanings of the word. The first defines it as a theory that states that reality is composed of a plurality of entities and the second defines pluralism as a state of society in which members of diverse ethnic, racial, religious or social groups maintain an autonomous participation in and development of their traditional culture or special interest within the confines of a common civilisation (Webster's New Collegiate Dictionary 1977).

The concept of pluralism is 70 years old and was defined only following the definition of <u>monism</u> (i.e. absolutism) and in opposition to it in view of the multifarious nature of reality. Pluralism is often used as a synonym for diversity and is frequently used in ethnic studies.

Dahl (1980) uses pluralism in two different but related ways. First, there is <u>conflictive pluralism</u> defined by the number and pattern of relatively enduring cleavages which characterise conflicts among persons. Second, there is <u>organisational pluralism</u> or the number and autonomy or organisations characterising conflicting situations among a given collection of persons. Cleavages which characterise conflictive pluralism are based on social classes, language, religion, race or ethnic groups and represent, perhaps, the most widely studied area of pluralism.

CONFLICTIVE PLURALISM

Many researchers claim that ethnicity is the most prominent factor in conflictive pluralism. Others give priority to class and socio-economic order in society. Many ethnic groups are traditionally mutually exclusive and solidarity within the groups is often based on common language, cultural history and a common future (Parsons 1975: 53-83; Douglas, this volume, Chapter 7). Cultural discord (referring to the underlying incompatibilities among fundamental beliefs held by different groups) is not a new theme in geographical enquiry (Evenden and Cunningham 1977: 7). Expressions of cultural discord include political conflicts, such as those rooted in disagreements over territorial claims, economic, religious and linguistic conflicts.

Discords and conflicts based on race are frequent themes among geographers and non-geographers alike. For instance, the racial conflict in South Africa can be analysed from several different perspectives. The apartheid policy of the South African government supports the formation of Black reserves and encourages the migration of Blacks to the newly formed Bantustans. Christopher (1982: 127-38) has pointed to the success of this policy; between 1970 and 1980, white South Africa became 'whiter'. The great dilemma of apartheid is that economic prosperity depends to a large degree on an integrated economy while the policy calls for racial segregation and the building up of the Bantustans (Pollock 1971: 364-67).

Race is the cornerstone in the Botswana state-idea of developing a multiracial state free of tensions as compared with the racial policy of South Africa (Knight 1977). Racial conflicts have been prominent in Southern Africa during the 1970s, with the guerrilla warfare in Rhodesia and mounting warfare in Namibia and against South Africa itself (Stephenson 1975). Moreover, the economy of states in Central and East Africa is strongly influenced by a reaction to the racial policy of South Africa (Hoyle 1978: 31-50; Reitsma 1974: 421-29). Race and ethnic identity of Blacks play an important role in American society (Morrill, this volume, Chapter 8). The electoral behaviour of Blacks shows that they often tend to give bloc support to candidates of their own race (O'Loughlin 1980). Similarly, recent riots in the United Kingdom have also had a racial background.

Second to race as a source of discord are linguistic differences. Linguistic discord seldom appears as the sole factor. Generally, it is accompanied by differences in religion and ethnicity. In Belgium, society is fragmented into the Flemish north and the Walloon south, but the north is also Catholic and is economically and politically inferior to the south (Claeys 1980: 169-89).

The religious component is highly influential in ethnicity, as the Canadian and Irish cases clearly indicate. The ethnic pluralism of Canada is transmitted into the political system through the federal government's implementation of bilingualism

and biculturalism and in the secessionist trends of Quebec
(Thorburn 1980: 151-68; Cartwright 1981: 205-24; Knight, Chapter
11). The Northern Ireland problem is often described erroneously
as a conflict between the Protestant and Roman Catholic communities
(Boal and Douglas 1982); it is more accurately perceived as
cultural and national discord (Pringle 1980; Boal 1980). Boal
and Douglas (1982) present Northern Ireland as a plural society,
through its cultural attributes, political, demographic and
economic behaviours.

Ethnicity is intimately related to the evolution of the
political systems in the nineteenth and twentieth centuries
through the force of nationalism. Ironically, just as ethnicity
and politics go hand in hand in the process of state- building,
it is also manifested in the expansion and fragmentation of the
twentieth century state system (Stack 1981: 3-15). Recent
disputes in Western Europe have been based on differences in
both race (England), language (Belgium), religion (Northern
Ireland, Netherlands, Germany). Many of the ethnic minorities
of Europe speak the language(s) of a neighbouring state and
problem areas such as Alsace, Upper Silesia and Schleswig are
on the frontiers of contiguous states. Irredentism plagued
Europe whenever ethnic minority groups were located just beyond
the limits of a state controlled by that ethnic group. Different
ideologies were also behind the various claims and even
anticolonialism can be interpreted as an ideological motivation
for a territorial claim (Burghardt 1973: 225-45).

Different territorial claims clash in the Israeli- occupied
West Bank. While it is still striving to integrate within Israeli
society, the Arab minority is still segregated to a large degree
from the rest of Israeli society (Soffer; Brawer; Rowley, this
volume, Chapters 9,13 and 15). The new nations of Africa face
an acute problem of integration as a result of their tribal
diversity. The new boundaries failed to coincide with ethnic
territorial boundaries (Parsons 1975: 53-83). Independence
movements keep the small Pacific Islands separate just as the
Caribbean states are independent and separate with little viability
as states (Haas 1970: 97-120). Sovereign states are sensitive
to the degrees of irredentism on their peripheries and dedicate
considerable effort towards 'nationalising' the peripheries.
Mexico attempted to increase the integration of border regions
within the national economy without losing the economic benefits
it enjoys from the U.S. (Dillman 1970: 243-48). Similarly, the
Dominican Republic removed all Haitian influence from their
mutual border (Augelli 1980).

At the micro level of cities, there are examples of cities
divided, such as Belfast (Boal and Livingstone, Chapter 12) and
cities which are integrated such as Jerusalem (Haupert 1971:
312-19). On the transnational level, we find that ethnic conflict
tends to influence international instability in a number of
regional subsystems. The most prominent cases are refugees such
as the Palestinian or South East Asian refugees and foreign

3

workers, who endanger the stability of some Western European states.

ORGANISATIONAL PLURALISM

Harold Laski was among the first to use the concept of pluralism in the area of politics. Federalism was a central doctrine of Laski's pluralist conception (Breitling 1980: 1-19). Federalism and federal systems were discussed from pluralistic perspectives by political scientists, sociologists and geographers. Federalism is considered to be the most geographically expressed form of government because it is based on regional differences and regional variations (Robinson 1961: 2). Federalism is a compromise between centripetal and centrifugal forces (Dikshit 1971: 97-116). Hartshorne specified certain centrifugal forces, such as physical barriers which handicap communication between regions, sparsely inhabited areas and areas inhabited by people ethnically different from the majority (Hartshorne 1950: 129-37). The failure of federalism in Central Africa is accounted for by the centrifugal force of the segregationalist policy of Southern Rhodesia (Dikshit 1971b: 224-29). McColl (1967; 1975) also seems to have adopted the theme of 'areally differentiated' pluralism in his work on the geographical bases of revolutions. The pillars of power in Asian revolutions are population, terrain, organisation and power, religion and race (McColl 1975: 301-10).

Organisational pluralism is determined, among other things, by the polyarchal nature of the regime, the structure of political institutions and the autonomy of various organisations within society.

Von Beyme (1980: 80-102) offers a check-list for pluralism in different spheres of society in which he includes the party system, the interest-group system, marginal and regional groups and pluralism in the mass media. Governments adjust the political systems of plural societies in order to answer for diversity and to prevent irredentism. Nigeria reorganised its areal government between 1967 and 1977 as a response to the dangerous pluralism in that country (Mawhood 1980: 103-16). Canada issued The Official Languages Act (1969) in order to determine bilingual districts (Cartwright 1981). Local government reform and reform in regional and administrative districts are also carried out with pluralism in mind (Burghardt; Honey; Paddison, Chapters 17,18 and 19).

PLURALISM, DEVELOPMENT, CAPITALISM AND SOCIALISM

The issue of development cuts across conflictive and organisational pluralism. Dahl (1980) has pointed to the possible cleavages formed by social classes in addition to language, religion, race or ethnic group. Radical geographers tend to

4

interpret the state of developing and developed regions as resulting from the organisational structure of the Capitalist system.

Dahl also considers the question of control as prior to the question of ownership of means of production, hence the question is not whether an order is socialist or non-socialist but how much autonomy is permitted to economic enterprises and the nature of internal and external control. Thus, a socialist economy can be highly decentralised and therefore pluralistic or highly centralised and monistic. Some observers claim that socialist economies are generally highly centralised, and thus less pluralistic, and this is especially true for socialist developing countries in the Third World. Others claim that the organisational structure of capitalist systems is based on transnational corporations which control the economies of Third World countries, therefore those countries are less pluralistic (see Cowie 1980; Susman 1981; Seidman and O'Keefe 1980).

Reitsma (1982: 125-30) argues that a division between developed and underdeveloped world (Ginsburg 1973) or between the First World (Capitalist Developed Countries), Second World (Socialist States) and Third World (The non-Socialist Underdeveloped Countries) is too simplistic. He recommends that we subdivide both capitalist and socialist countries into developed and underdeveloped and this system of four types of country allows ten types of external relationships. The socio-economic order of developing regions and developing states are highly influenced by transnational entities. Foreign labour and foreign investment in Third World states provide good examples.

Thus, Cowie shows that the political economy of colonial Zambia was based on its incorporation into South Africa's regional economy, the development of its mining industry and its classification as a non-settlers territory. These three processes designed by the colonial power determined the future interaction network (Cowie 1980: 47-54). Reitsma (this volume, Chapter 20) has pointed to the dependence of developing African land- locked states on their neighbours for free passage of people and commodities and how this dependence shapes their political and economic networks. Particular attention was focused on the role of transnational corporations which have been responsible for the industrialisation of subcentres in South Africa using cheap, and abused, Black labour (Susman 1981; Seidman and O'Keefe 1980). Ethnic, political and state of development identities project into transnational relations especially in the U.N. and other international organisations. Abate and Brunn describe the emergence of African voting blocs and alliances at the U.N. and Brunn and Ingalls (this volume, Chapter 21) have found that this applies for the Soviet Union and her Eastern European allies.

SUMMARY

We end this Introduction with the conclusion that political geographers initially reached the study of pluralism from the perspective of organisational pluralism and only recently have turned to conflictive pluralism. This path is similar to that followed in the development process of other social sciences such as Political Science and Sociology. A relatively new issue which has attracted both traditional and non-traditional (radical) geographers is that of development and pluralism. This is probably the most _political_ topic nowadays among political geographers and will most likely prove to be a focus for much research in the coming decade.

REFERENCES

ABATE, Y. and S.D. BRUNN [1977]
    'The emergence of African voting blocs and alliances in the U.N., 1961-70', Professional Geographer, 24, 338-46.
AUGELLI, J.P. [1980]
    'Nationalization of Dominican borderlands', Geographical Review, 70, 19-35.
BOAL, F.W. [1980]
    'Two nations in Ireland', Antipode, 12(1), 38-44.
BOAL, F.W. and J.N.H. DOUGLAS (eds.) [1982]
    Integration and Division: Geographical Perspectives on the Northern Ireland Problem, (Academic Press, London).
BREITLING, R. [1980]
    'The concept of pluralism', in S. Ehrlich and G. Wootton (eds.), Three Faces of Pluralism, (Gower Publishing Co., Farnborough), 1-19.
BURGHARDT, A.F. [1973]
    'The bases of territorial claims', Geographical Review, 63, 225-45.
CARTWRIGHT, D. [1981]
    'Language policy and political organization of territory - A Canadian dilemma', Canadian Geographer, 25, 205-24.
CHRISTOPHER, A.J. [1982]
    'Partition and population in South Africa', Geographical Review, 72, 127-38.
CLAEYS, P.H. [1980]
    'Political pluralism and linguistic cleavage: the Belgian case', in S. Ehrlich and G. Wootton (eds.), Three Faces, 169-89.
COWIE, W.J. [1980]
    'Articulation, global capitalism and the spatial organisation of colonial society', Antipode, 12(2), 47-54.

DAHL, R. [1980]
'Pluralism revisited', in S. Ehrlich and G. Wootton (eds.),
Three Faces, 20-33.
DIKSHIT, R. [1971a]
'Geography and federalism', Annals, Association of American
Geographers, 61, 97-116.
DIKSHIT, R. [1971b]
'The failure of federalism in Central Africa - a
politico-geographical post-mortem', Professional Geographer,
23, 224-29.
DILLMAN, D.C. [1970]
'Recent development in Mexico's national border program',
Professional Geographer, 22, 243-48.
EVENDEN, L.J. and F.F. CUNNINGHAM (eds.) [1977]
Cultural Discord in the Modern World, (Tantalus Research,
Vancouver).
GINSBURG, N. [1973]
'From colonialism to national development - Geographical
perspectives on patterns and policies', Annals, Association
of American Geographers, 63, 1-21.
HAAS, A. [1970]
'Independence movements in the South Pacific', Pacific
Viewpoint, 97-120.
HARTSHORNE, R. [1950]
'The functional approach in political geography', Annals,
Association of American Geographers, 40, 95-130.
HAUPERT, J.S. [1971]
'Jerusalem: aspects of reunification and integration',
Professional Geographer, 23, 312-19.
HOYLE, B.S. [1978]
'African politics and port expansion at Dar-es-Salaam',
Geographical Review, 68, 31-50.
KNIGHT, D.B. [1977]
'Racism and reaction: the development of a Botswana "raison
d'etre" for the country', in L.J. Evenden and F.F. Cunningham
(eds.), Cultural Discord, 111-26.
McCOLL, R.W. [1967]
'A political geography of revolution: China, Vietnam and
Thailand', Journal of Conflict Resolution, 2, 153-67.
McCOLL, R.W. [1975]
'Geographical themes in contemporary Asian revolutions',
Geographical Review, 65, 301-310.
MAWHOOD, P. [1980]
'The government of Nigeria: structural change as a response
to pluralism', in S. Ehrlich and G. Wootton (eds.), Three
Faces, 103-16.
O'LOUGHLIN, J. [1980]
'The election of Black mayors', Annals, Association of
American Geographers, 70, 353-70.

PARSONS, T. [1975]
'Some theoretical considerations on the nature and trends of change of ethnicity', in N. Glazer and D.P. Moynihan (eds.), Ethnicity - Theory and Practice, (Harvard University Press, Cambridge), 53-83.

POLLOCK, N.C. [1971]
'The dilemma of apartheid in South Africa', Tijdschrift voor Economische en Sociale Geografie, 62, 364-67.

PRINGLE, D.C. [1980]
'The Northern Ireland conflict - a framework for discussion', Antipode, 12(1), 28-38.

REITSMA, H.A. [1974]
'Malawi's problem of allegiance', Tijdschrift voor Economische en Sociale Geografie, 65, 421-29.

REITSMA, H.A. [1982]
'Development geography, dependency relations and the capitalist scapegoat', Professional Geographer, 34, 125-30.

ROBINSON, K.W. [1960]
'Sixty years of federation in Australia', Geographical Review, 50, 1-20.

SEIDMAN, A. and P. O'KEEFE [1980]
'The United States and South Africa in the changing international division of labor', Antipode, 12(2), 1-16.

STACK, J.F. (ed.) [1981]
Ethnic Identities in a Transnational World, (Greenwood Press, Westport).

STEPHENSON, G.V. [1975]
'The impact of international economic sanctions on the internal viability of Rhodesia', Geographical Review, 65, 377-89.

SUSMAN, P. [1981]
'Regional restructuring and transnational corporations', Antipode, 13(2), 15-24.

THORBURN, H.G. [1980]
'Ethnic pluralism in Canada', in S. Ehrlich and G. Wootton (eds.), Three Faces, 151-68.

VON BEYNE, K. [1980]
'The politics of limited pluralism? The case of West Germany' in S. Ehrlich and G. Wootton, Three Faces, 80-102.

8

PART I: THEORY AND POLITICAL GEOGRAPHY

Chapter 2
THE QUESTION OF THEORY IN POLITICAL GEOGRAPHY

Peter J. Taylor

GROWTH AND CONFLICT IN POLITICAL GEOGRAPHY

There has been an increasing awareness in recent years of the importance of 'the political dimension' in human geography research. This has produced two distinct trends; (i) existing fields of enquiry have become more political in nature and (ii) there has been a large growth in the traditional field of political geography. Hence we are all familiar with the fact that urban, economic and social geography now incorporate political themes into their subject matter at the same time that political geography itself is said to be experiencing a renaissance. While these alternative arrangements for accommodating the newly perceived importance of the political into human geography are themselves a fascinating topic, producing as it were political Geographers and Political Geographers, I will limit my consideration to the latter in this paper.

Although research on traditional themes in political geography has continued at a steady pace – Cohen (1982) on geopolitical structures, House (1981) on boundaries, Gottmann (1980) on the spatial structure of the state, Fifer (1981) on a capital city – the real growth in the subject has come from researchers exploring new themes. With the advent of these 'new' political geographers has come academic conflict as they have expressed their dissatisfaction with the heritage of traditional political geography. Johnston (1980), for instance describes a 'political geography without politics' as being 'weakly developed within the current corpus of human geographical work because its practitioners ignore the real content of politics.' The preface of Cox's (1979) new textbook for political geography is similarly provocative:

> 'A preface provides an author with the opportunity to set forth the concerns which inform his project. In this instance the concerns are especially clear: a dissatisfaction with the state of political geography realised

over a decade of attempts to develop a course in the topic for undergraduates. Four problems have seemed central: internal coherence; external coherence; scale; and relevance. On each of these criteria considerable shortcomings have been evident.'

Reynolds (1981) is more succinct but perhaps even more damning when he begins a paper with the simple assertion that 'Political geography is theory-deficient'. It is the purpose of this short essay to try and clarify the nature of these academic conflicts with a view to their resolution through mutual co-existence.

MACPHERSON'S QUESTION AND ANSWER

My starting point for this task is a paper by Macpherson in which he addressed his fellow political scientists with the question 'Do We Need a Theory of the State?'. This was prompted by the dominance of empirical theories of political processes in modern political studies which have eclipsed interest in more general theories of the state. Whereas theories of political process may provide adequate descriptions of the activities of modern governments they make no attempt to link political organisation to 'supposed essentially human purposes and capacities' which is the hallmark of 'grand theories'. Macpherson's question is who needs these latter theories (Macpherson 1977)?

Macpherson's answer is, in many ways, even more interesting than his question. Basically he concludes that for the constituency of political scientists some do need such a theory and some do not! He is able to reach this conclusion by dividing up his constituency into three categories.

1. Those who accept and uphold the existing democratic- liberal state with only minor reservations. This includes the bulk of the empirical theorists and some 'philosophic liberal' theorists. This category has dominated modern political science especially in USA.

2. Those who accept the normative values of late nineteenth century liberal theorists (J. S. Mill) but feel that the current state does not realise those values. This includes the bulk of non-Marxist socialists and social democrats.

3. Those who do not accept the existing state and social order and who reject theories that uphold or reform it. They would replace both by Marxian theory and practice. We can label these categories the <u>status quo</u> group, <u>reformers</u> and <u>radicals</u> respectively.

Macpherson argues that the status quo group have no need of a theory of the state, the reformers do need one but have not got one and the radicals not only need one but have also been actively producing such a theory in recent years. Macpherson writes from a reformist position and the bulk of his paper illustrates how reformers can and should learn from recent neo-Marxist literature on the theory of the state.

The first point to notice about Macpherson's argument is that by dividing up the constituency he is able to define different needs. Simply recognising this fact can prevent exhortations from one group to another falling on deaf ears. This is how I interpret much of the recent academic conflict in political geography. Johnston's (1980) aim is wide of the mark when he castigates traditional political geography for not having a theory of the state. Quite simply such a theory was not necessary to their needs. Traditional political geography is only 'theory-deficient' on terms laid down by 'new' political geographers.

This argument suggests that we can apply Macpherson's categories to the constituency of political geographers. Of course the different institutional contexts and histories of the political science and political geography constituencies, not least in terms of their relative sizes, will make any exact correspondence impossible. Nevertheless I think it is possible to categorise political geographers into the three types, status quo group, reformers and radicals. The critics quoted at the beginning of this paper are reformers attempting to break away from political geography's strong status quo heritage. They have now been joined by Marxist geographers who are incorporating political geography into their broader political-economy perspective. However since it is the 'middle' reformist group that have been largely responsible for the recent growth in political geography I will follow Macpherson and concentrate on their needs in the next section. I will then diverge from Macpherson's discussion in subsequent sections since my personal position is from a radical stance.

POLITICAL PROCESS AND THE STATE IN POLITICAL GEOGRAPHY

The recent growth in political geography was largely initiated by concern to incorporate the political process in geographical research. This led to geographers borrowing the empirical theories available in political science notably for studying elections and urban planning issues. After a short honeymoon with political science many of these researchers began to develop doubts about these theories similar to those expressed by Macpherson. A theory of the state was required within which these particular research topics could be set.

The first presentation of theories of the state in geography from a non-Marxist perspective was that of Dear and Clark (1978) in which Marxist ideas were expressed but the repertoire of political science was maintained. This paper tends to confuse functions of the state with theories and this enables the authors to combine a wide range of models and theories in a single framework. Their 'conceptual interdependencies between five theories of the state' has Marxist theories of the state as an over-arching theory within which the non-Marxist ideas of political scientists are accommodated.

Dear and Clark's approach has been followed by Johnston (1980) although he more carefully distinguishes function and theory and provides us with six theories in all. He spells out the relationship between the Marxist theory and the others:

> 'The framework for the explanation of any particular event or pattern by political geographers is provided by the state-as-an-associate-of-capitalism theory. But this cannot account for the particular event/pattern, it merely establishes the degrees of freedom, the limits to individual behaviour. Explanation of the particular requires the use of other theories, such as that of the pluralist-decision-maker.' (Johnston 1980: 445)

To summarise: The Marxist theory of the state is necessary but not sufficient for a full explanation; non-Marxist theories are required to cover for its supposed deficiencies.

Finally we need to mention Hall's (1982) recent discussion where he directly compares his own 'political science' approach to Castell's (1977) Marxist analysis. His discussion is interesting because he relates these theoretical positions to actual research activities concerning very similar case studies. He compares his own study of the planning decisions concerning motorway development around London in the late 1960s\early 1970s to Castell's work on Paris motorway development for the same period.

In each case an initial consensus for motorways (planners, politicians, public) gave way to a political reaction against the plans. In London the plans were stopped; in Paris the change did not come quickly enough to prevent their construction. Hall's study of the London situation employed a social choice framework on how the decisions came to be changed and, in particular, how politicians and planners reacted to the change in public opinion. However he admits that he was not able to tackle adequately the question as to why public opinion changed. In contrast Castell's Marxist interpretation of the Paris situation is, according to Hall, much better able to deal with changing public opinion as part of the unfolding economic crisis of the recession but he is weak on how this public opinion is translated into policies. Hence Hall (1982: 75) is able to conclude that:

> 'The two traditions could offer a great deal to each other, since their level of theoretical explanation is somewhat different and quite possibly complementary.'

Once again the answer seems to be obvious - put the theories together and a more complete explanation may be possible.

Such a convenient conclusion is only valid if the two traditions are indeed complementary. I will argue that you cannot pick and choose pieces of theory to fit together an overall explanation any more than a data analyst can pick and choose those sets of information which match his particular hypothesis. Marxist theory is much more than a set of constraints and the political scientist's theories of political process have assumptions that extend well beyond simple decision-making. The two traditions incorporate two sets of assumptions that are incompatible. At the most fundamental level they incorporate opposite theories of human purposes and capacities - aggressive, competitive individuals versus creative, cooperative individuals. Different pieces of the two theoretical traditions may fit together at a rather superficial level of explanation but they can never be integrated in the way Hall (1982: 75) seems to envisage when he looks 'hopefully to the development of some eclectic body of theory'. Quite simply the same individuals being modelled in a vote-buying exercise and who operate under constraints defined by a structural Marxist theory cannot be essentially co- operative and competitive in nature at the same time: The theories that Dear and Clark, Johnston and Hall attempt to combine are incompatible.

THE 'POLITICS' IN POLITICAL GEOGRAPHY

At this point of the discussion I am going to diverge from Macpherson's argument and consider further the 'politics' of the three constituencies we have identified in political geography.

Stripped of their academic masks, the traditional status quo group in political geography are ultimately concerned with order. It is for this reason that they do not need a theory of the state: the order imposed by the current state system is accepted as given. Traditional political geography has always been about order expressed in spatial structures as currently illustrated by Cohen's geopolitical models concerned with international spatial structure and order between states and Gottmann's core-periphery models concerned with intranational spatial structure and order within states.

Two examples from two different generations will illustrate this point. Goblet (1955: 20) provides a very clear statement of this particular politics of political geographers. He argues that the political geographer has a 'practical aim' and a 'scientific goal'. In its practical aim political geographers provide an 'information service' for governments and their diplomatic corps. The second aim is an empirically-based search for geographical laws of the rise and fall of states in which he sees the twentieth century as 'one large laboratory' for

13

political geographers. This empirical bias is again very explicit in the recent work of House (1981) on the USA-Mexico boundary. He terms his approach 'applied' which he defines as taking 'an operational stance, perhaps on occasion involving service to a client' (House 1981: 291) and involving particular concern to avoid ideological biases. 'For this reason the method advanced is empirically based', House (1981: 291) continues. The result of his efforts is an abstract model of boundary interactions which may have a role in controlling the situation along the border; the phrase he uses is 'tension management'. Of course boundary studies in political geography epitomise the concern for spatial order with the origins for this interest in the division of non-European areas among imperialist states in the seminal work of Holdich (1916).

As we have seen the reformist group of 'new' political geographers did not criticise their traditional 'foes' for their conservative politics but emphasised their supposed theoretical inadequacies. Nevertheless the reformist group brought with them into political geography a new package of topics for study. The key word changes from order to welfare. This is most explicit in Cox's (1979) textbook where political geography is equated with welfare geography but in general this group are concerned with distributional issues within a welfare state context. In one sense this represents a distinct break with the traditional political geographers in that they refute Hartshorne's (1950) advice to concentrate on regional integration at the expense of 'vertical' (i.e. socio-economic) integration. From another perspective, however, it can be argued (Taylor 1982: 17) that the new emphasis upon welfare issues is merely a belated attempt to bring political geography into line with the phenomenal growth of state activities in the twentieth century: The rise of welfare states requires the rise of welfare political geography.

It is these 'new' political geographers that have been the borrowers of ideas from political science as they have attempted to make the subdiscipline more 'modern'. It is ironic therefore that they are borrowing from the status quo group within the political science constituency. The product of such a process can only be a new 'modern' status quo group to replace the old 'traditional' status quo group. This result is most clearly seen in the use made of the social choice paradigm in recent years in political geography (Archer 1981). When Reynolds (1981) attempts to correct theory-deficient political geography he develops an argument for minimal state interference in the nineteenth century liberal tradition. He employs Rawl's theory of justice which Macpherson identifies as part of the 'philosophic liberal' component of political science's status quo group. Hence we can look forward to the emergence of a status quo group in political geography much closer to their equivalent group in political science including both an 'empirical theory' component and a 'philosophic liberal' component.

Reynolds' work is the most explicit in using the individualist assumptions of the social choice approach but these ideas are found in the work of other 'welfare' political geographers (Hall 1974; Cox 1979; Johnston 1980). It is interesting, however, that these other members of the reformist group are less sure of the efficacy of these new models. Like Macpherson they seem to want to avoid the full implications of working in a framework associated politically with the 'new Right'. According to Macpherson (1977: 226) this liberal 'model of man and society is becoming morally repugnant to increasing numbers of people within the liberal democracies, as well as in the world at large'. He wrote this before the Thatcher and Reagan electoral successes so we may doubt the generality of the statement but it is certainly true for many reformists, hence Macpherson's own consideration of Marxist theories of the state. A similar route is being taken by some reformist political geographers notably Dear (1981), Clark (1981) and Cox (1983).

What is left of the 'new' political geography that emerged in the 1970s therefore? Clearly the reformist group is splitting as some turn right (Reynolds, Hall) and others left (Dear, Clark, Cox, Johnston?) leaving the centre unoccupied. Their efforts to combine the incompatible may be seen as a brave attempt to generate a viable alternative to the status quo and radical positions. It has been unsuccessful because there can be only two types of theory: traditional and critical. The former support the status quo, the latter expose it as part of a process of overthrowing (Lewis & Melville 1978). The reformers we have discussed have had to choose which way their researches will point, right or left. The end result of the process will be two basic political geographies although each will express a variety of related positions within an overall perspective.

I have discussed at some length alternative status quo political geographies but what of this critical alternative, a radical political geography? I will not go into details here since I have recently expressed some ideas on this topic elsewhere (Taylor 1982). The following points are worthy of note, however. Most radical geography has been political geography with a small 'p'. The particular status of a subdiscipline is not an important issue in the broad, holistic political economy approach. Nevertheless the recent emphasis upon theories of the state in political economy has meant that political geography can become a convenient label to attach to some of this research. It is ironic that this approach has the theoretical apparatus to integrate the themes of both control and welfare and hence transcend the conflict with which we began this paper. Political geography within political economy has the potential for becoming an important vehicle for studying the interface of space and power from a critical perspective (Hudson 1982; Dear 1982).

DEBATE OR LEARNING?

The conclusion that there can be just two fundamental political geographies provides some interesting implications. First we should avoid debates where we argue past each other. The conflict which I used to open this essay is a case in point. It is no good exhorting people to develop a theory of the state when they do not need one. There can be debate <u>within</u> each perspective but since the goals are diametrically opposed, debate <u>between</u> the two political geographies would seem to be pointless. In the end this will merely reduce to 'exposing' each other's ultimate aims. I propose that we begin at this point. Accept that these opposite goals exist and proceed from there.

But where do we proceed? The poorer route is to ignore each other as both perspectives are developed independently. However much some would support this, it is not really a viable alternative. The most productive way forward is to learn from one another; even to use one another. Although I have criticised Hall's (1982: 75) notion of an 'eclectic body of theory' I fully endorse his call for 'some mutual understanding'. The empirical tradition of the status quo group, for all its faults, provides, and will continue to provide, a wealth of information for critical theorists to evaluate. Similarly, following Macpherson, the non-radical researchers have much to learn from the insights provided by the recent upsurge of neo-Marxist writing.

I will conclude with a comment on the group I have been most unkind to in the essay, the reformist political geographers. As Macpherson (1977) admits for political scientists, they have not been able to develop a distinctive position in the twentieth century. I think the same is true for my constituency of political geographers and the onus is squarely on their shoulders to prove me wrong by their deeds. As we begin to appreciate more and more the appalling effects of the capitalist world-economy for 'Third World' peoples, it begins to make the reformist attempt to delineate a 'nice' capitalism look like nothing more than their particular version of the search for the holy grail.

NOTES

1. The largest group of political Geographers are, of course, the radical school whose journal <u>Antipode</u> provides very 'political' articles on urban and economic geography. There are interesting contrasts outside this school, however. In the early 1970s, Wolpert, Mumphrey and Seley (1972) and Cox (1973) were both studying land use conflicts within the cities but only the latter designated himself a Political Geographer. Similarly, in the late 1970s, Smith (1977) and Cox (1979) both developed a welfare geography, the former as a new human geography, the latter as a new political geography.

2.   Macpherson admits that the taxonomy is not all-inclusive
but it does cover the main lines of thought in modern debate. I
will use this argument to justify my taxonomy also.

REFERENCES

ARCHER, J.C. [1981]
    'Public choice paradigms in political geography', in A.D.
    Burnett and P.J. Taylor (eds.), Political Studies From
    Spatial Perspectives, (Wiley, Chichester), 73-90.
CASTELLS, M. [1977]
    The Urban Question, (Edward Arnold, London).
CLARK, G.L. [1981]
    'Democracy and the capitalist state', in A.D..Burnett and
    P.J. Taylor (eds.), Political Studies, 111-130.
COHEN, S.B. [1982]
    'American foreign policy and global geopolitical equilibrium',
    Political Geography Quarterly, 1, 223-41.
COX, K.R. [1973]
    Conflict, Power and Politics in the City: a Geographic
    View, (McGraw-Hill, New York).
COX, K.R. [1979]
    Location and Public Problems, (Maaroufa, Chicago).
COX, K.R. [1983]
    'Residential mobility, neighbourhood activism and neighbourhood
    problems', Political Geography Quarterly,
    (forthcoming).
DEAR, M. [1981]
    'A theory of the local state', in A.D. Burnett and P.J.
    Taylor (eds.), Political Studies,
DEAR, M. [1982]
    'Research agendas in political geography: a minority view',
    Political Geography Quarterly, 1, 178-79.
DEAR, M. and G.L. CLARK [1978]
    'The state and geographic process: a critical review',
    Environment and Planning A, 10, 173-83.
FIFER, J.V. [1981]
    'Washington D.C.: the political geography of a federal
    capital', Journal of American Studies, 15, 5-26.
GOBLET, Y. [1955]
    Political Geography and the World Map, (George Philip,
    London).
GOTTMANN, J. (ed.) [1980]
    Center and Periphery, (Sage, Beverly Hills).
HALL, P. [1974]
    'The new political geography', Transactions, Institute of
    British Geographers, 63, 48-52.

HALL, P. [1982]
   'The new political geography: seven years on', Political
   Geography Quarterly, 1, 65-76.
HARTSHORNE, R. [1950]
   'The functional approach in political geography', Annals
   of the Association of American Geography, 40, 95-130.
HOLDRICH, T.H. [1916]
   Political Frontiers and Boundary Making, (Macmillan,
   London).
HOUSE, J.W. [1981]
   'Frontier studies: an applied approach', in A.D. Burnett
   and P.J. Taylor (eds.), Political Studies,
HUDSON, R. [1982]
   'Political geography and political economy', Political
   Geography Quarterly, 1, 178-79.
JOHNSTON, R.J. [1980]
   'Political geography without politics', Progress in Human
   Geography, 4, 439-46.
LEWIS, J. and B. MELVILLE [1978]
   'The politics of epistemology in regional science', in
   P.W.J. Batey (ed.), Theory and Method in Urban and Regional
   Analysis, (Pion, London), 82-100.
MACPHERSON, C.B. [1977]
   'Do we need a theory of the state?', Archives of European
   Sociology, 18, 223-44.
REYNOLDS, D.R. [1981]
   'The geography of social choice', in A.D. Burnett and P.J.
   Taylor (eds.), Political Studies, 91-110.
SMITH, D.M. [1977]
   Human Geography: a Welfare Approach, (Edward Arnold,
   London).
TAYLOR, P.J. [1982]
   'A materialist framework for political geography',
   Transactions, Institute of British Geographers, N.S. 7,
   15-34.
WOLPERT, J., A. MUMPHREY and J. SELEY [1972]
   Metropolitan Neighborhoods: Participation and Conflict
   over Change, Resource Paper #16, (Association of American
   Geographers, Washington D.C.).

Chapter 3
THEORY AND TRADITIONAL POLITICAL GEOGRAPHY

Saul B. Cohen

There are those who believe in supporting the current state system, those who believe in its overthrow, and those who are undecided. That most Political Geographers fall into the first category is hardly surprising. The large majority of the people who live on this earth, including residents of Communist countries, adhere to the principle that the State is the best available mechanism for maximising the cultural and economic potential of individuals and groups. They believe that control of territory by the State mechanism is indispensable to their security, welfare and life-satisfactions.

To suggest, as does Peter Taylor, that 'traditional' Political Geographers do not have or need a grand theory of the State, but that Marxists do, is a dubious proposition. Moreover, he has defined these traditionalists as 'status quo empiricists' who accept the order of the State system as a given. But surely the nature of the order makes for profound differences in the relationships between a government and its peoples. Options include: centralisation of structure versus decentralisation; homogeneity versus heterogeneity of population; the role of the State either as reconciler of different ethnic and class interests, or as the mechanism that imposes the will of a particular group – majority or minority – on the rest; the dominance of society versus the preeminence of the individual.

Also, to describe traditionalists as empiricists is to imply that most of us in this category take the empiric-deductive approach, in contradistinction to the radicals and the middle-of-the-roaders who presumably represent the theoretical- deductive approach. That there are many empiricists among the traditionalists is true. But many traditional Political Geographers also take the theoretical-deductive route.

As one who has been strongly influenced by my mentor Derwent Whittlesey and by Stephen Jones and Jean Gottmann, I have used theory as the starting point in my work (Gottmann 1973; Jones 1954; 1955). No theory is value-free. Regardless of geographical scale, the ideas and values of those who control the body politic and its spatial structures and landscapes, are at the roots of political-geographical analysis. Moreover, Political Geographers who are apolitical and who are content with sterile analysis of form, simply have rejected the theoretical-deductive approaches espoused by the field's founding fathers.

Ratzel's organismic state, Mackinder's global spatial equilibrium, Bowman's prescriptions for international order, and Whittlesey's ideal state provide mainstream Political Geographers with inspiration for theory-building that represents a very different tradition from that of the succeeding functionalist approaches which concentrated on how the State works (Ratzel 1897; Whittlesey 1944; Mackinder 1919; Bowman 1921; 1931). The writings of each of the above reflect deep concern with ideology and social order, as part of the focus on the interplay between politics and nature.

For Ratzel, the nation's need to fulfil its mission as a dominant culture could be satisfied only by a continuing process of spatial growth in the context of Darwinian theory. His was an organismic theory of political dominance and subordination. Mackinder had a firm belief that the freedom of men was linked to the freedom of nations. His vision was that of an interdependent world in which the preservation of capitalistic values required some measure of control over world economic growth. To preserve Western freedom, a global balance could be maintained only by checking Heartland's expansion into the maritime reaches of World Island. In essence, Mackinder espoused a spatial balance-of-power theory.

Bowman was interested above all in a world of international cooperation and order. Reduction of armaments, enlightened colonialism, creating such instruments as the League of Nations, and the use of land for social needs were policies that he espoused. While recognising the need to protect minorities within the State, he was a vigorous spokesman for the creation of new national boundaries that would embrace national ethnic groups and reduce minorities within those boundaries. Bowman embraced a theory of conflict resolution based upon international and national consensus.

Whittlesey believed that inhabitants of every distinctive physical region had the impulse to create political frameworks around them. This urge to coalesce was the basis for the dynamic interplay between government and nature, and historic ties to the land supported this urge. Out of such thinking emerged his concepts about the ideal size and shape of the State, and the role of core-peripheral relations in maintaining the body politic. Whittlesey's theory was that of the harmony of the system.

All of these founding fathers of traditional Political Geography were as 'political' in their approaches, as are today's radicals and reformers. As a traditional Political Geographer who has tried to follow the mainstream approach, I have an ideological view of the State as a reconciler of diverse cultural interests, and a manager and regulator of the distribution of scarce economic resources. Because I believe that national self-sufficiency is neither possible nor desirable, I support national policies that promote proximate regional cooperation and global interdependence. Because I feel that societies, like individuals, are systems that evolve in predictable ways, I try to apply organismic-developmental theory to the behaviour of the national and the international system. Geopolitical systems, regardless of scale, are dynamic. To maintain their equilibrium they must remain hierarchical, specialised and integrated - open to outside forces and absorbing these forces in a way that permits evolutionary changes. All of these beliefs are grounded in political theory and value sets. The shortcomings of my work stem not so much from lack of theory as from empiric evidence.

Just as traditional Political Geography cannot be apolitical, it cannot be aterritorial. Traditional Political Geographers assume an unending cycle of interrelationship between politics and space. It matters not whether a people's initial values were shaped by a very special type of environment, or whether a people with pre-set values shaped the environment to help them apply those values to everyday life. Once the cycle begins, a feedback process carries it along (Cohen 1976).

Radical geographers and reformers, many of whom have British experiences, may be more heavily influenced by their own national class-conscious environment than they care to admit. The strength and persistence of working-class neighbourhoods and values in the United Kingdom fortifies their theories of historical materialism. However, most traditional Political Geographers do not assume that economics and class are the only, or even the essential basis for linking a people with territory. Indeed, the evidence throughout much of the world is that race, religion and culture are more important operational forces in the politics of space at all geographical scales, than is economics. In fact, class struggle is increasingly out-of-date as a concept, not simply in the Western world, but in the socialist world. Rather, cultural struggle is the main basis for domestic and international strife. National states have become more heterogeneous, and inter-national conflict has increasingly revolved around issues of religion, race, ethnicity, politics and history - not resources.

With the advent of modern telecommunications and high technology, it was assumed that particularised space would become less important as a vehicle for forging socio-cultural values, and that people would maintain desired networks through telecommunications. This has not proven to be the case, because politics are rooted in space. Regardless of whether the national electoral structure of the State encourages localism, the tendency of culture groups to cling to neighbourhoods, cities, rural areas and regions remains a fact of political life.

Thus, in America, minorities cling to certain sections of cities for the political power that they derive from their territorial base. The political struggle is far less frequently a coalition of the poor against the rich --it is rather framed in socio-cultural terms - in terms of age, of sex, of religion, of race and of ethnicity. Blacks fight against gentrification of their neighbourhoods by middle class Whites; the poor of one ethnic group resist moves into public housing of other poor; the elderly oppose entry of young couples into suburbs of restrictive zoning that prohibits lower-cost multiple family dwellings; Jews, Greeks and Italians encourage co-religionists to move into neighbourhoods already anchored by synagogues and churches. In the developing world, the rural peasantry uses violence and political clout to maintain its way of life; and ethnic minorities persistently cling to regions that constitute their folk-fortresses.

Many traditional Political Geographers are cultural determinists, just as radicals are economic determinists. The former are no less theoretical than the latter. Moreover, many traditionalists who hold a positive theory of the State are no less theoretical than radicals or reformers who have a negative or ambivalent theory of the State. Perhaps Peter Taylor should have said that amongst Political Geographers there are non-Marxists, Marxists and those of indeterminate views, that within all three categories there are those who in practice use the theoretical-deductive and those who use the empiric-inductive approaches. This would have more accurately reflected the situation.

Irrespective of these criticisms of the categorisation, by focusing the topic as he has done, Peter Taylor has performed a service for traditionalists by challenging them to search anew for a theoretical basis in their works. Moreover, he has reminded all who are interested in the field, regardless of label, that too much of what has been written in sterile empiricism that has neither emerged from theory nor led to the building of theory. Rather it has focused on the description of phenomena that either is served up in a political vacuum or is presented as an extension of political slogans.

REFERENCES

BOWMAN, I. [1921]
   The New World, (World Book Company, Yonkers).
BOWMAN, I. [1931]
   The Pioneer Fringe, (American Geographical Society, New York).
COHEN, S.B. [1976]
   'Political landscape and value system: a developmental approach', (Indian Geographical Society, Jubilee Volume).

GOTTMANN, J. [1973]
   The Significance of Territory, (University of Virginia
   Press, Charlottesville).
JONES, S.B. [1954]
   'A unified field theory of political geography', Annals,
   Association of American Geographers, 44, 111-23.
JONES, S.B. [1955]
   'Views of the political world', Geographical Review, 53,
   309-29.
MACKINDER, H. [1919]
   Democratic Ideals and Reality, (Holt, New York).
RATZEL, F. [1897]
   Politische Geographie
WHITTLESEY, D. [1944]
   The Earth and the State, (Published for the United States
   Armed Forces Institute, Madison by H. Holt).

Chapter 4
WHO NEEDS THEORY? A RESPONSE FROM THE SCHIZOPHRENIC MIDDLE GROUND

Ronald J. Johnston

Taylor's adaptation of Macpherson's essay to the context of political geography has produced a stimulating and argumentative piece. The classification on which the argument is based is a useful summary of contemporary political geography, and I am happy to respond as a leftward-leaning, middle-ground reformer. I can, of course, speak only for myself and in no way implicate others in what I have to say.

A PRELIMINARY: WHAT IS THEORY?

Both Macpherson and Taylor conclude that their first category of scholars - status quo empiricists - do not need a theory of the state (Macpherson talks of 'grand theory' (Macpherson 1977)). I find this somewhat surprising, since as far as I am concerned, all work, however close to the empiricist ideal (which is unattainable), must be theory led. I believe that the members of the 'status quo' group have a theory; so, it seems, does Pete Taylor in his statement that 'the order imposed by the current state system is accepted as given'.

Clearly it is necessary for me to provide a working definition of theory. For this, I draw upon Harvey (1969: 88), who presents a theory as a series of statements that provide a framework for the consideration of particular topics, in this case the empirical issues studied by political geographers. It is, then, an organisational framework which structures how one thinks about relevant subject matter. It is akin to an ideology, in the positive sense of that word (Johnston 1982a), except that a theory is used by a (social) scientist to guide empirical investigations and can be amended by objective testing.

Given this definition I find some of the Macpherson/Taylor argument unconvincing. Any social scientific work must be theory-led (however poorly articulated the theory). Whether a group needs a theory is irrelevant; what is relevant is the nature of the theory it is using. In this context, I argued (Johnston 1980a; 1981) that the theory of state used by political geographers in Taylor's 'status quo' grouping is deficient. To my mind, this is because its view of the state is false, mainly because the nature of that crucial body has not been examined critically. The only published response (Chisholm 1981) does not contest this basic point.

Elsewhere, Peter Taylor has noted that I entered political geography from urban studies and have no roots in the subdiscipline (Burnett and Taylor 1981: 7); leading me, it seems, to dismiss 'as irrelevant the past heritage of the field' (Taylor 1982: 33). I was 'particularly dismissive' not because the subject-matter of his Type I political geography is irrelevant but because the way it is studied does not illuminate the questions I ask regarding 'why?'. Such a situation continues: Gerald Blake (1982) has rightly pointed to the interesting contemporary issues of maritime sovereignty, but addresses no questions as to why this is currently such an important issue and why the state is involved. It is for this reason that I sought a reconstituted theoretical position.

PROBLEMS IN THE MIDDLE GROUND

The position that I have taken (Johnston 1982b) appears untenable to Peter Taylor. I have tried to marry the grand theory of the state as derived from Marxist work with a perspective on the role of the individual. This does not imply that 'The Marxist theory of the state is necessary but not sufficient for a full explanation', hence the search for remedies to its 'supposed deficiencies'. To me, Marxist theory is sufficient for what it purports to do but is insufficient for my objectives.

To make my position clear, the following is a brief resume.
1.   Geography is an empirical discipline, concerned with particular events, patterns and phenomena. It seeks to understand why and how they come about (Johnston 1980b; 1982c). In political geography, this implies a focus on empirical issues with a political content defined as those involving the active participation of the state.
2.   The Marxist theory of the state, developed by the Type III scholars in the Macpherson/Taylor classification, provides an excellent rationale for the state and for its activity in two linked areas - support for capitalist accumulation (particularly for those capitals that are based within the state's territory) and societal legitimation of the capitalist system. But, as Marxist scholars have made clear this theory cannot account for empirical events - their details of time, place and substance. Marxist theory is a theory of society. As Eyles (1981) suggests, without it, work in human geography is likely to be deficient. But if geographers want to study empirical aspects of society, they must develop a theory of state action within that Marxist context.

3.   Decisions within capitalist society are taken by knowing individuals. They do not change the structure of that society (now a single world-economy as Taylor (1982) has argued persuasively), although they may deflect its trajectory slightly - hence Giddens (1981) concept of structuration. But they do produce its empirical realities. They are responsible for what is done where, and when, and to explain such outcomes the Marxist theory is not irrelevant but insufficient: it is the framework - 'both enabling and constraining' according to Giddens - within which individuals act.

An empirical discipline, therefore, needs both a general theory of society and a modus operandi for looking at the detailed outcomes: an empirical political geography needs a grand theory of the state and a framework for studying state actions.

I have aimed in recent years to develop such a modus operandi, a way of studying how individuals act and produce particular geographies (Johnston 1982b). The subject matter may well include that of traditional political geography. Peter Hall has been working in a similar eclectic way, as illustrated by his theoretical discussion of Great Planning Disasters (Hall, 1981).

Peter Taylor believes that this modus operandi is not viable, because it seeks to marry traditions that are not complementary but incompatible. Clearly I disagree (see also Cox and Johnston 1982). According to Taylor, this is because the Marxist theories and the idealist theories that I seek to combine 'incorporate opposite theories of human purposes and capacities'. The one (Marxist) sees people as aggressive and competitive and the other as creative and cooperative. I would answer this in two ways. First, the two theories are not opposite because cooperation may be necessary, at least in the short run, to achieve one's competitive goals. Secondly, being creative, aggressive and competitive are not necessarily in opposition. And in any case, even if we accept that we are all aggressive/competitive, the related instincts come out differently in different people and produce geographies. What I am interested in is who wins, what they do, and what the outcome looks like: Marxist theory only tells me why they are competing.

THE USES OF POLITICAL GEOGRAPHY

Peter Taylor claims that the middle-of-the-road reformers are caught between the 'support the state' political position of the status quo group and the 'overthrow the state' view of the radicals. There are three issues here.
1.   First, do you achieve change by either imposing it or arguing successfully for it? I cannot countenance the first, hence my adherence to the second. As Taylor points out, the latter is the goal of critical theory. Pedagogically, the theoretical element of that position needs empirical support as illustration, hence my work (Johnston 1982b; Cox and Johnston 1982).

2.   Secondly, what change? On this I admit a dilemma. My work convinces me that capitalism is a bad, if not an evil, system and that reform from within will achieve no removal of the bad elements. I want it changed. But to what? I don't know; I see no viable alternative. But this does not stop me wanting to expose my understanding of the capitalist system - in my empirical way - to as many people as will listen to and read what I have to say, and perhaps then discover a sensible alternative.

3.   Thirdly, might I inhibit change? Possible; all knowledge is potentially dangerous, as Harvey (1973) made clear in his discussion of counter-revolutionary theory. It may well be that what I do is counter-revolutionary, because it provides the capitalist system with tools for further repression. But most theory, however revolutionary in intent, is potentially usable in that way, if its proponents believe in argument not force. Thus Taylor is right that my search for a 'holy grail' may be misguided. I can only plead honesty - and hope.

Clearly, in this response, I have emphasised the points of disagreement, for this is the purpose of debate. Peter Taylor and I agree on many things. I believe his essay's clarity has highlighted what to me as well as to him are the fundamental issues of contemporary political geography.

REFERENCES

BLAKE, G.H. [1982]
    'Maritime boundaries and political geography in the 1980s',
    Political Geography Quarterly, 1, 170-74.
BURNETT, A.D. and P.J. TAYLOR (eds.) [1981]
    Political Studies from Spatial Perspectives, (Wiley,
    Chichester).
CHISHOLM, M. [1981]
    'Political geography without prejudice', Progress in Human
    Geography, 5, 593-94.
COX, K.R. and R.J. JOHNSTON (eds.) [1982]
    Conflict, Politics and the Urban Scene, (Longman,
    London).
EYLES, J. [1981]
    'Why geography cannot be Marxist: towards an understanding
    of lived experience', Environment and Planning A, 13,
    1371-88.
GIDDENS, A. [1981]
    A Contemporary Critique of Historical Materialism, (Macmillan,
    London).
HALL, P. [1981]
    Great Planning Disasters, (Penguin, Harmondsworth).

HARVEY, D.W. [1969]
   Explanation in Geography, (Edward Arnold, London).
HARVEY, D.W. [1973]
   Social Justice and the City, (Edward Arnold, London).
JOHNSTON, R.J. [1980a]
   'Political geography without politics', Progress in Human
   Geography, 4, 439-46.
JOHNSTON, R.J. [1980b]
   'On the nature of explanation in human geography',
   Transactions, Institute of British Geographers, N.S. 5,
   402-12.
JOHNSTON, R.J. [1981]
   'British political geography since Mackinder', in A.D.
   Burnett and P.J. Taylor (eds.), Political Studies, 11-
   32.
JOHNSTON, R.J. [1982a]
   Approaches to Human Geography, (Edward Arnold, London).
JOHNSTON, R.J. [1982b]
   Geography and the State, (Macmillan, London).
JOHNSTON, R.J. [1982c]
   'On the nature of human geography', Transactions, Institute
   of British Geographers, N.S. 7, 123-25.
MACPHERSON, C.B. [1977]
   'Do we need a theory of the state?', Archives of European
   Sociology, 18, 223-44.
TAYLOR, P.J. [1982]
   'A materialist framework for political geography',
   Transactions, Institute of British Geographers, N.S. 7,
   15-34.

Chapter 5
THE QUESTION OF THEORY IN POLITICAL GEOGRAPHY: OUTLINES FOR A CRITICAL THEORY APPROACH

Ray Hudson

Peter Taylor draws a distinction between 'Political Geographers' and 'political geographers'. As I would not claim to be a 'Political Geographer' but rather, like all geographers, 'political', I do not feel competent to comment upon the disputes within 'Political Geography' beyond making a couple of simple points. First, the identification of 'Political Geography' as a discrete sub-discipline and the resulting claims made for it, particularly the assertion that it is non-ideological, perform an important ideological function by attempting to depoliticise politics. Second, the assumptions made within 'Political Geography' as to the nature of both 'the Political' and 'Geography' are of central importance. How, for example, is the political process to be conceptualised - in terms of the geography of voting patterns or in terms of the structure of class relations, most fundamentally those of production? Clearly most 'Political Geographers' would not wish to be associated with the latter view which makes its 'political' stance explicitly clear.

Taylor concludes that, fundamentally, there are two political geographies based on different conceptions of theory, one associated with 'critical theory', the other with 'traditional theory', that these have diametrically opposed goals, and that we should accept this situation and proceed from there - while recognising that acceptance of the concept of a 'critical theory' poses problems. This is nevertheless a position that I would accept not just for 'political geographies' but more generally for human geography (Hudson 1981). Given Taylor's emphasis upon theorising about the State, a fundamental issue for both 'Political' and 'political' geographers, I shall develop some of the implications in terms of what sort of historical-materialist theories of the State are needed to explore within the context of a critical theory approach. This may well constitute fresh territory for many 'Political Geographers'.

It is important to stress at the outset that, within an historical-materialist approach, there is no question of there being a 'general, universally valid theory of the state'; to believe this would be to miss a fundamental point about historical materialism. Rather the point at issue is to specify particular types of theory relating to particular types of State. To give a simple example, different theories would be required to deal with the issues of the absolutist State (Anderson 1974) and the capitalist State; the problems posed by the State that forms part of 'the alternative in Eastern Europe' (Bahro 1978) requires yet another theoretical response. At another level, the capitalist State itself has changed profoundly through time and this development has demanded parallel developments in theory. The State in the United Kingdom in 1800 was very different to the contemporary UK State (see Nairn 1977). The contemporary UK State differs from other European States (see Scase 1980) while there are considerable differences between it and the contemporary Brazilian State, for example (Munck 1979).

It is this issue of adequately theorising the capitalist State that I wish to examine in greater depth. It seems to me that 'political geographers' have increasingly become concerned with it in their attempts to break the mould of the old 'Political Geography'.

Jessop (1979) specifies five criteria against which an historical-materialist theory of the capitalist State should be judged. First, it must be grounded in the specific qualities of the capitalist mode of production, the requirements imposed by surplus-value production and the accumulation of capital. Second, it must therefore assign class struggle a central place in the accumulation process, given its decisive role in determining the rate of surplus-value. Third, it must establish and specify the relations existing between the political and economic without mechanistically reducing one to the other. Fourth, it must allow for historical and national differences in the forms and functions of capitalist States. Fifth, it must allow for the influence of non-capitalist class and non-class forces in determining the nature and extent of State power within a given social formation. What these latter three criteria make clear is that there cannot be one universally valid historical-materialist theory of the State; rather different theories have been developed in different times and places which reflect the character and evolution of States as the capitalist mode of production itself and social formations under its sway have developed. A failure to recognise this can lead to misdirected critiques of particular theories. Simmie (1981: 85-86), for example, dismisses all historical-materialist theories on the grounds that some of them, such as the 'classical' texts of Engels, Gramsci, Lenin, Marx, Trotsky, fail to deal with situations that developed long after the deaths of their creators. Yet to expect them to do so is not only unreasonable but reveals a fundamental failure to grasp the nature of historical-materialist theory.

Equally, however, it would be a mistake to believe that more recent theories of the capitalist State are unproblematic and that one can, as it were, merely take one ready-made off the shelves, for difficulties exist with them on several levels.

Different theories tend to emphasise different aspects of the capitalist State - and hence play down others. This can be illustrated, drawing on Jessop's discussion and the five criteria set out above, by reference to the emergence of the 'capital logic' school's conception of the State as the ideal collective capitalist and the theoretical responses to this view. The capital logic school, centred on the Free University of Berlin, attempted to derive the general form and principal functions of the capitalist state from a pure capitalist mode of production and its conditions of existence. It argued that the separation of State and civil society is not only possible but necessary, as an institution not immediately subordinated to the disciplines of the law of value and competition through the market, is needed to provide general pre-conditions that cannot or will not be provided by individual capitals. Insofar as the common needs of 'capital in general' are met by a distinct political institution - the State - then the State acts as an ideal collective capitalist, though one that is still subordinated to the laws of motion of capital. While apparently leaving the capitalist State firmly enmeshed in the contradictions of the capitalist mode of production, this approach nevertheless represents a theoretical advance controlled by capital (see Poulantzas 1973). Rather, it is an essential element in the social reproduction of capital and as such there can be no simple relationship between the interest of capital and State intervention, given that capital exists both at the level of 'capital in general' and at that of 'many (competing) capitals', hence the seemingly muddled nature of State policies (see Hudson and Lewis 1982; Hudson 1983; Mohan 1983).

In other respects, however, the capital logic approach raises serious difficulties. Fundamentally, it asserts the necessity of a political level - the State - the forms and effects of which are determined at the economic level. Without it, capitalism is, as it were, impossible. This, however, merely points to capitalism as a possible mode of production, with a specific associated form of State. It suggests possible or probable forms of the capitalist State, specifying broad structural limits within which it could develop but not why particular capitalist States evolve in particular ways within them.

To begin to answer this latter question involves introducing historical specificity and the role of class struggle. An important contribution was made to this by a group of theorists centred on Frankfurt (see Holloway and Piciotto 1978). One can also cite Offe's (1975) work here, for while starting from different assumptions, he reached broadly the same conclusions. Central to their response is a rejection of an analysis of the State simply in relation to the needs of competing capitals and a recognition of the importance of the antagonistic relation between capital and wage-labour. Thus emphasis is placed upon the need to understand the capitalist State in terms of its changing function in class struggle over the organisation of the labour-process and the production and appropriation of surplus-value. Consequently, attention is directed to the historical development of this struggle rather than to the logical implications of the existence of many competing capitals and to the changing forms of State intervention that become appropriate as the accumulation process itself develops. This focus upon the historical unfolding of the accumulation process leads to considerable weight being attached to the role of crises in capitalist development and to State intervention being interpreted in terms of crisis- avoidance and crisis-management strategies.

Such a view, then, represents an advance while preserving the main theoretical gains yielded by the capital logic approach; of utmost importance, it attempts to show that it is impossible for the State to secure all the needs of all capitals at a specific time. Equally, though, it has its limitations. For example, no recognition is given to the influence of social forces and classes other than those of capital and wage- labour in class struggles. Such limitations are particularly apparent in transitional periods. Abstract analysis of the capital-State-wage-labour trilogy cannot resolve the problems posed by issues such as the changing relationships between the feudal nobility and bourgeoisie in the transition from feudalism to capitalism, nor indeed that transition itself, nor can it explain the effects of, say, religious ideology in Northern Ireland (see Farrell 1976). Yet these remain crucial issues in understanding the nature of the State power in particular societies and, therefore, its relationship to the accumulation of capital.

The concept of ideology is central in the writings of Gramsci (1971) and the 'neo-Gramscian' school. In particular, attention is focused upon the concepts of political and ideological hegemony – the imposition of a particular view of the social world by a ruling class and its acceptance by subordinate classes, consonant with the interests of the former. The role of the State in unifying the ruling class, in organising its ideological and political hegemony is seen as decisive. Unity depends upon particular forms of organisation and representation of class interests. Class practices, in terms of organisation and representation, become the critical element in securing conditions for continuing accumulation. The dominant class must be organised, subordinate classes disorganised. The problems posed by this organisation and disorganisation are to be resolved via the nature of ideological hegemony and/or the form of the State. Dominated classes and important social groups organised on non-class lines must come to accept the legitimacy of the power bloc's 'view of the world'. It must be emphasised that this does not merely represent a state of 'false consciousness' on the part of these classes and groups; rather it comes about as a result of the incorporation of some of their aspirations and interests into the dominant ideology. One important channel through which this can come about is that of parliamentary democracy, the formulation of reformist social welfare programmes which meet the demands of groups outside the power bloc at the same time as they satisfy the requirements of accumulation (see Hudson 1983).

In a sense we have now come full circle. For if a weakness of the capital-logic school was that it allowed no place for political practice and class struggle, then a weakness of the 'class theoretical', Gramscian approach is that it can lead to an underestimation of the structural constraints which the capitalist mode of production imposes on the scope for autonomous State action and an over-estimation of the relative autonomy of ideology and politics. Often, economic problems (for example, of profitability) cannot be solved at the same time as socially-progressive reformist packages are pushed through. The sacrifices of the latter may call into question the hegemonic position of the ruling bloc and the legitimacy of the State (see Habermas 1976). Put another way, we may end up in a situation where bourgeois democracy cannot secure the conditions for continuing accumulation. For a critical theorist what happens next, of course, is not amenable to prediction in the sense in which a traditionalist theorist uses that term.

What I have attempted to demonstrate is that to adopt a 'critical theory' approach to the problems of theorising the State constitutes a considerable advance but raises fundamental problems - in both a theoretical and political sense. What emerges, though, as central to elaborating an adequate theory of capitalist States is the relationship between agency and structure, between human actions and thoughts and the social structural relations in which these are enmeshed and within which they have to be located.

In this light, Johnston's comment to the effect that it is necessary to graft on non-Marxist to Marxist theory to deal with the deficiencies of the latter misses the point by a considerable distance and is based upon a partial understanding of historical materialism, confusing and equating it with a rather arid structuralism. This is a point that cannot be over- emphasised for it means that Johnston and others of his persuasion fundamentally fail to appreciate the central analytic thrust of historical materialism. It is correct to identify different levels of analysis but equally incorrect to suggest that these have to be dealt with separately within different theoretical frameworks, and then welded together; historical materialism itself provides the necessary methodological framework for such multi-level analyses. The challenge for those who take up Peter Taylor's call to develop an alternative to 'Political Geography' that draws its inspiration from critical theory is to demonstrate the viability of this project via concrete analyses of particular situations which use such a framework, confronting theories with the evidence of the social world and thereby elaborating the former while helping bring about a progressive transformation of the latter.

REFERENCES

ANDERSON, P. [1974]
  Lineages of the Absolutist State, (New Left Books, London).
BAHRO, R. [1978]
  The Alternative in Eastern Europe, (New Left Books, London).
FARRELL, M. [1976]
  Northern Ireland: The Orange State, (Pluto, London).
GRAMSCI, A. [1971]
  Selection from the Prison Notebooks, (Lawrence and Wishart, London).
HABERMAS, J. [1976]
  Legitimation Crisis, (Heinemann, London).
HOLLOWAY, J. and A. PICIOTTO (eds.) [1978]
  State and Capital: a Marxist Debate, (Edward Arnold, London).
HUDSON, R. and J. LEWIS (eds.) [1982]
  Regional Planning in Europe, (Pion, London).
HUDSON, R. [1981]
  'State policies and changing transport networks: the case of post-war Britain', in A.D. Burnett and P.J. Taylor (eds.), Political Studies from Spatial Perspectives, (Wiley, Chichester), 467-88.
HUDSON, R. [1983]
  'Capital accumulation and regional problems: a study of North-east England 1945-80', in F.E.I. Hamilton and G. Linge (eds.), Spatial Analysis, Industry and the Industrial Environment, Volume 3, (Wiley, London).
JESSOP, B. [1979]
  'Recent theories of the capitalist state', Cambridge Journal of Economics, 1, 353-73.
MOHAN, J. [1983]
  State Policies and Public Facility Location: the Hospital Service of North East England 1948-82, (Ph.D thesis, Department of Geography, University of Durham), (forthcoming).
MUNCK, R. [1979]
  'State and capital in dependent social formations: the Brazilian case', Capital and Class, 8, 34-53.
NAIRN, T. [1977]
  The Break-up of Britain: Crisis and Neo-nationalism, (New Left Books, London).
OFFE, C. [1975]
  'The theory of the capitalist state and the problem of policy formation', in L.N. Lindberg, R. Alford, C. Crouch and C. Offe (eds.), Stress and Contradiction in Modern Capitalism, (D.C. Heath, Farnborough).

POULANTZAS, N. [1973]
   Political Power and Social Classes, (New Left Books, London).
SCASE, R. (ed.) [1980]
   The State in Western Europe, (Croom Helm, London).
SIMMIE, J. [1981]
   Power, Property and Corporatism, (Macmillan, London).

PART II: PEOPLE — HUMAN PERSPECTIVES ON PLURALISM

Chapter 6
EQUITY AND FREEDOM IN POLITICAL GEOGRAPHY
Paul Claval

INTRODUCTION

Since the 1950s, new directions in political geography have developed partly under the aegis of a new political sociology. Systems analysis has supplied a fresh vision of the political process. The State appeared as a machine and its function was to respond to pressures from the population. The whole structure of the nation was conceived in terms of information, feed- backs and inputs and outputs of the political subsystem. The new political geography of the 1960s drew heavily upon this view of government. There was soon some disillusion since it was hard to define a spatial dimension to a political circuit. It is difficult to evaluate information flows when no distinction is made among them, when it is practically impossible to develop the idea of the range of power as economic geography has developed the idea of the range of a good or when casual news of no direct political significance is undifferentiated from information relevant to political actors, i.e. from orders and control. At the beginning of the 1970s, the new political geography had to face a difficult situation. The orientation that looked most promising ten years earlier proved to be difficult to pursue. It was certainly possible to develop it, but for that to happen, attention to the diversity of information flows in the political system was essential. But few geographers understood this situation (Claval 1979).

For the majority, there were two possibilities. First, the problems of war and peace and the energy crisis invited a return to the study of the State and geostrategy. Second, the radical orientation of social sciences during the 1970s gave strength to the study of the geographical impact of political action within the State. The main question raised concerned spatial equality and the corresponding question of spatial equity.

To address the question of social justice, the intellectual equipment of the time seemed sturdy and versatile. Rawls (1971) had recently emphasised the real possibilities offered to everyone with procedural justice as a prerequisite but not a sufficient

36

condition for a fair, equitable society. Geographers were invited
to stress the economic aspect of the equity problem. In modern
industrial societies, large differences have persisted between
the incomes of white- and blue-collar workers. In urban settings,
some people have good access to services while others have low
incomes and poor accessibility to jobs and to commercial or
service centres.

Much of the new political geography of the 1970s was built
along these lines, building on developments in economic geography.
The problems tackled by many radical geographers have nothing
to do with the exercise of power in space, focussing instead on
central places, range of goods and on questions about the provision
of public goods by the local, regional or national government.
The results of this new orientation are quite impressive (Burnett
and Taylor 1981), but they seem peripheral to the development
of political geography.

In the study of equity problems in space, some consideration
has been given to the spatial characteristics of certain types
of political processes. Spillover effects are ubiquitous in the
cities. As a result of neighbourhood effects, people can influence
one another. Poor tenants may be a detriment to wealthy landowners
who cannot be repaid. Whinston and Davies, followed by geographers
such as Harvey (1973) and Cox (1973; 1978) have shown the
prevalence of this kind of diffuse power effects in the development
of the complex settlement structure of modern metropolises. But
nothing was done to extend this approach to other political
processes in the city or in the State.

THE INDIVIDUAL EXPERIENCE OF FREEDOM AND POLITICAL GEOGRAPHY

In Britain and in North America, the radical movement in
the social sciences emphasised the problem of social justice and
equality. A similar situation developed in Scandinavia, the
Netherlands and to some degree, in Germany. In Latin- speaking
countries, there is also a radical trend in the social sciences,
but equality is only one objective of study. The influence of
Marxism was noticeable after World War II in both Italy and
France. In the 1950s, for instance, French geography was dominated
by members of the Communist Party who were eager to denounce all
forms of inequality in Western societies. Since then, there has
been an increasing interest in social justice. However, what
is more significant today than the question of inequality is
that of freedom. Geographers know that Eastern European societies
have achieved some measure of success in their social policies;
at the same time, practically all traditional liberties have
disappeared. A similar evolution is seen as possible elsewhere,
hence the emphasis on civil rights.

The term 'freedom' has individual as well as social meanings.
At the personal level, questions persist as to whether there is
anything like free will, or are people influenced by heredity,

history or environment? At the collective level, is it really possible for people to express what they wish? Although the second question is more geographical than the former, we shall show that both are relevant for modern political geography.

The economic pre-conditions of freedom

In the social sense, to be free is to be independent, to be able to choose without interference from others. Freedom has a basic economic dimension. Political economists have been well aware of this fact since classical times. In the eighteenth century, Thomas Jefferson preferred rural development for America since for him the real basis of democracy was the existence of a class of self-sustaining farmers.

It appears that unequal conditions exist for freedom in urban and rural areas. As long as there was a frontier, younger sons of farmers could move to the new lands. There was no need for subdivision of landholdings with each new generation. But the situation has become quite different with the arrival of the finite world. Pierre Chaunu (1974) has shown that in Western Europe, the transition between the open world of the early Middle Ages to the filled-up world of the later Middle Ages and modern times is a major divide. With strong pressures on the main resource base - land for much of the history of humankind - it became more difficult to provide everyone with the reality and the sense of economic independence without which there is no possibility of choice.

As the city evolved, its trading functions grew more important. Its economic base ceased to be in the form of rent and land taxation, as the skills of merchants and craftsmen developed. With the increasing division of labour, cities became more and more the place for freedom.

But with the oncome of capitalism, things changed. It is difficult to speak of an economic independence for the manual workers of the mills in the early nineteenth century. As strong trades unions began to develop towards the end of that century, the situation changed. Blue-collar incomes began a steady rise in Western Europe and the United States. Yet, even with high wages, wage earners remained dependent, in the sense that there is no security for them. The possible loss of a job is still a shocking reality.

The economic basis for freedom represents a classical study. Marxists have devoted a good part of their efforts for over a century towards showing the harsh limitations of choice for the majority of the population even in more affluent societies. That there is a trade-off between freedom and equality is obvious to many. Societies, like the Swedish, which emphasise the same achievements for all, have had to pay for this by reducing the possibilities for personal initiative. We shall not develop this well-known point. Our purpose is to show that there is a threshold below which one cannot speak of real freedom while above it, the trade-off is less fundamental than people generally think. What is more important is the way different kinds of freedom are made accessible.

Freedom and accessibility

To ensure monetary incomes and security of employment is no guarantee that freedom will follow. This lesson is certainly one of the conspicuous features of Eastern European states where people are always certain of being employed but have few real freedoms.

To be free, in a second meaning of the word, is to be able to make choices. It implies being confronted by many opportunities. The more complex the situation and the more diversified the society, the better the position of the decision-maker. Freedom is an essential feature of urbanisation. With more activities in a city, it becomes easier for people to find precisely what they seek.

For social scientists interested in spatial equality, the main problem is the provision of evenly distributed central places that provide easy access to services to all. As soon as a person can access services rapidly planners deem everything to be all right. However, people often prefer some choice among different locations and different services. Spatial uniformity is undoubtedly a better solution for equity than for freedom. And, as long as people are not provided with cheap information on everything, the best solution is for them to live in a large city if they want to be really free. Such freedom is the meaning of the slogan of Henri Lefebvre 'le droit a la ville' (Lefebvre 1968).

Telecommunications and mass media continue to transform steadily as we live through a sociological urbanisation of the world (Remy and Voye 1974). In this age of the dispersed city, freedom is certainly less easily identified with the metropolis than it was half a century ago, but for the more exceptional components of freedom, New York, San Francisco, London or Paris retain a strong appeal.

Space and role competition

It seems, at first glance, that the exercise of free will has practically nothing to do with geographical realities, but on further reflection the conclusion is different. People are freer when they have more scope and better information. Other aspects of the decision process are also geographically relevant. Decision makers are not merely utility maximisers - or satisficers. They are persons and as such, they are involved in a perpetual self-building process. When conflicts over values arise, their solution implies choices between sets of rules. It is more difficult to cope with difficulties when you have to be perfectly coherent. When it is possible to play different roles at different times, it is possible to choose easier ways and to reach the same aim more safely. But it is impossible to do so when dealing always with the same people. If you are well-known by all, your freedom of choice is restricted by general expectations. People will perhaps help, but you must be faithful to them and to yourself if you desire their support. Public control is strong in small societies and range of choice is narrow in restricted environments.

In a large city, a different situation exists. You can adopt other attitudes when your partners change: they have different ideas and expectations of you. Anonymity produces freedom. Social control is less efficient when you associate with different people.

Goffman (1959) has developed good insights into the ways in which people use different environments to play out their roles. They are not always on the stage. It is important for them to have places to escape public attention, to recollect their favourite memories, to think or to dream. In a city, it is easier to live behind the scenes than is the case in a village. Sansot (1978) and Remy (Remy and Voye 1980) point to a contrast between primary and secondary spaces. The first are those of daily life, of family, of work. The second are places where it is possible to avoid people, to establish relationships selectively and to escape neighbours. Jean Remy is particularly interested in second homes in the life of French-speaking bourgeoisie in France and Belgium. The double morphology is a means of achieving a combination of roles more satisfactorily and of performing cherished private roles that are incompatible with most professional and civic responsibilities.

Free choice is certainly not a product of the physical setting in which people live, but certain spatial organisations are more propitious than others for expressing personal preferences or idiosyncracies.

Freedom and security

The freedom of everyone is limited by the freedom of others. It is impossible to analyse the spatial problem linked to the use of liberty when working only at the level of the isolated individual. Obstacles always arise between the intentions and actions of participants on the social stage. The first effect of these interrelations is negative – social control that is adverse to freedom. But it can help it.

To be really free, it is important to avoid relations with authoritarian or violent characters. The control exercised by the collectivity on them is good for all those unable to achieve the same degree of independence by themselves. As long as there are no institutions specialising in social regulation, no police, no legal rules which define everyone's rights, the solution for having some measure of freedom is to rely on the local community and its power to oversee everything.

When people live without written laws or without a judicial system, freedom is tied to the primary group. A person has more real opportunities in a small village than in a city ruled by a minority of aristocrats or wealthy merchants.

The modern city offers a good environment for expressing personal preferences and achieving self-realisation as long as security does not present a problem. With the disappearance of community control, it becomes more difficult to cope with delinquency and deviance. An hour spent at late night entertainment in the central city can be pleasant, but this is so only as long as you are not threatened by others. Ambiguity is an important attribute of urban atmosphere (Remy and Voye 1980).

In the nineteenth century city or in the early twentieth century, the strong class structure in Western European cities allowed for the coexistence of different people in the same environment. The present situation is different, especially in American cities where class structure has always been weaker and is today totally negligible in the field of social control.

Societies in large cities today are neither regulated by the collective reciprocal protection of the local community nor by the written law of society. The cities certainly offer good opportunities for certain types of people, generally young and unfettered by moral principles, but they can be ominous for the majority of the population. There is no general freedom in the jungle and large cities increasingly resemble a jungle.

The micro-scale analysis of the exercise of freedom is important for geographers. Most of the traditional studies by political scientists or sociologists focus on the problems of the nation, of the city or of the neighbourhood and not with the experience of individuals. Is the equity objective achieved if a country's citizens all enjoy access to all services? Apparently, yes, but in fact, no, because of the qualitative differences between the urban centres, people feel freer in the more complex environments.

To develop liberty, the best solution is certainly not uniformity. As long as the costs of information remain high, spatial concentration in big cities pays. The urban setting is also propitious to a freer composition of roles. Double morphology, certainly expensive for individuals and for society, is conducive to good conditions for the exercise of freedom.

The search for liberty as experienced at the microscale certainly implies strong differentiations in the spatial organisation of societies. Let us move then to the more classical problem of freedom at the societal scale.

FREEDOM AND THE CIVIL SOCIETY

People sometimes speak of political geography as if its main task were to study State and public administration. Since the eighteenth century, political philosophers have been aware that the regulation of social life is not achieved solely through the State. Civil society plays an important role in the control and adjustment of behaviour. Some mechanisms, like the market mechanisms, do not imply the use of power by individuals – but this is not true of societal relations built on a pyramidal or hierarchical principle: family, patronage, feudalism, caste and bureaucratic organisations all convey some form of power and collaborate with the State in the ruling of society as a whole. The exercise of freedom can only be understood in the global frame of the social architecture of the group. Since each type of societal relation implies transfers of a different mixture of news, moral rules, order and controls, its spatial characteristics are different (Claval 1979). The ranges of the relations are short or long depending on their content. The number of people encompassed in the system depends also on the nature of the hierarchical structure.

The forms of civil society differ widely among civilisations. Power is exercised in two ways: 1) as a collective control, as in neighbourhood relations or in inter-group contacts of caste societies and 2) as a hierarchical and interpersonal structure of news, orders and control transfer in families, in feudal or in pure political relations. The second solution is more efficient since it allows for a more adequate division of responsibilities, defining more precisely the limits of freedom for all participants. The system of societal relations, linked as it is with the system of property rights, maps out the nature and the field of choice devolved to all the members of a society.

The hierarchical structures of traditional societies are generally unable to rule over large groups of people, especially when they cover large areas. Sometimes the extension of the hierarchy is limited as in the familial system. Sometimes the efficiency of relations declines as soon as news and orders have to travel over more than two hierarchical rungs. Whenever they reject the power structure as illegitimate control costs are so high that distance is prejudicial to the smooth running of the system. In traditional societies, civil society cannot regulate large groups since its hierarchical structures are relatively inefficient. Order and peace are maintained through local collective control or through competitive equilibrium as in the balanced system of segmentary societies (Evans-Pritchard 1937) or in the market system of the Melanesian primitives (Pospisil 1963). In such societies, freedom as lack of order is important, but the reciprocal control is so tight that real possibilities of initiative are only open on the margin of the humanised space, on the frontier, or in no-man's-lands (Clastres 1974; Bonnemaison 1979).

Development of modern societies is linked with the emergence and growing role of organisations - either political, religious, military or industrial and commercial. Since their authority is legitimate, the costs of control are lessened. There are no logical limits to the encompassing power of the system and distances are easier to deal with.

The structure of freedom differs in modern societies as compared to traditional. Collective control is structurally minimised in hierarchies. Supervision by foremen, directors and overseers is always limited in its scope, and division of space allows for the exercise of restricted freedom at each level. Efficient organisations do not suppress freedom, but the range of initiatives is constricted. In such societies, liberty is not akin to anarchy, but to initiative and responsibility. When bureaucracies are built on a competitive basis, as in the market economy of modern societies, rigidity of control is always balanced by the search for flexibility and efficiency, so that freedom devoted to all is important even if it is unevenly distributed.

In industrial societies, freedom of everday life depends upon the possibilities of role composition opened by the multiplicity of situations and of partners in the large cities, and upon the opportunities of self-expression offered by professional life, by intellectual or artistic activities or by involvement in associations. The amount of liberty really available is important, but its manifestations are always fragmented. Where the organisations are more heavy-handed, the situation is often rather unpleasant.

The experience of freedom of the majority of the population depends not only upon political institutions, but reflects the distribution of settlement (for the part of collective control) and the role of bureaucracies in the economic and social life of the country. The distribution of freedom is unequal. Large cities are favoured because of their anonymity and because of the concentration of big firms or administrations. When insecurity threatens people in central cities, their suburbs become the more attractive locations.

The development of private bureaucracies in all the sectors of economic life certainly gives more opportunities of choice to many people, but these opportunities are restricted to a narrow field of personal ability. For daily life, the sense of freedom is certainly better preserved in more traditional societies where it is possible for everyone to participate in a wider range of activities and of responsibilities. This is certainly the reason why so many young people disagree with the social structures of industrial societies and try to change them, even if they are built on truly democratic political institutions.

FREEDOM AND THE POLITICAL PROCESS (STRICTO SENSU)

The political dimension of freedom is just a part of the whole experience of choice and self determination that can be provided in all societies. In democratic States, it implies the right to participate in the political process for everyone. It is easy to provide everyone with the elementary and fundamental right of expression everywhere through the electoral process, but this right is only a part of what is necessary for a wholesome political freedom. It is necessary in large modern societies for people with the same views, the same ideologies or the same interests to build associations if they want to express their positions and exercise influence on the political decision process. In the modern polyarchies (Dahl 1971), pressure groups and lobbies are sound and necessary institutions. They convey from bottom to top of the State structure more complex information than the 'yes or no' indications of the votes. The real freedom to participate in political life is less evenly distributed than it appears at first glance. Large centres of communication offer more facilities for ambitious people.

The political process in democratic societies also depends on the topology of sources of legitimation. In the formal Western democracies, there is a structural division between the power system and the right to decide what is good and what is bad for man and for society. The roots of this differentiation are as old as the distinction between temporal and spiritual power in mediaeval Europe, but its manifestations changed when the Church ceased to exercise a monopoly in the legitimation process. The emergence of public opinion in modern Europe is a sign of this transformation. Well-to-do people, especially the successful traders or businessmen, were the first to ask for the right to express their views on all fundamental questions. In fact, political freedom in this sense is generally the privilege of a small class of newsmen and of intellectuals. Potentially, everyone is able to write or to express his views on society. In fact, people who by their professional activities become well acquainted with the principles and subtleties of law, or manipulate collective symbols and build social ideologies, are the main opinion builders and the main legitimisers at the national level. As a result, the exercise of liberty of opinion is neither egalitarian in the social sense (it is a quasi-monopoly of small groups) nor in the spatial sense (there are generally only a few leading cities which compete in this process).

The exercise of political freedom does not imply a real uniformisation of space. Political circuits have nodal points where the main themes of political competition are elaborated, where decisions are made and where things happen.

CONCLUSION

The study of the geographical conditions of liberty in social life is conducive to a view of the spatial dimensions of the political processes that is much richer than that deduced from the analysis of social justice problems. In this sense, the conceptual analysis of the different forms of freedom is a necessary step for the emergence of a renewed political geography.

At the beginning of this century, modern sociology was built by people like Max Weber, Emile Durkheim or Wilfredo Pareto who devised a coherent and subtle system to describe and explain social life. At the same time, geographers thought of themselves as naturalists and failed to provide such a framework for their discipline. At the end of the 1950s, the first phase of development of a new geography consisted in borrowing from economics such concepts as transportation costs, externalities, economies of scale and range of goods. Sociology and political science had nothing equivalent to offer. The idea of systems analysis was certainly more promising but was insufficient when not propped up by adequate categories for analysing social and political life and processes. I think that we must now build this framework. The study of the geographical conditions of freedom appears to me to be one of the most exciting fields for such an intellectual venture.

REFERENCES

BONNEMAISON, J. [1979]
    'Les voyages et l'enracinement. Formes de fixation et de mobilite dans les societes traditionelles des Nouvelles Hebrides', L'Espace Geographique, 8, 303-18.
BURNETT, A.D. and P.J. TAYLOR (eds.) [1981]
    Political Studies From Spatial Perspectives, (Wiley, New York).
CHAUNU, P. [1974]
    L'histoire, science sociale, (SEDES, Paris).
CLASTRES, P. [1974]
    La societe contre l'Etat, (Editions de Minuit, Paris).
CLAVAL, P. [1979]
    Espace et Pouvoir, (PUF, Paris).
COX, K.R. [1973]
    Conflict, Power and Politics in the City: a Geographic View, (McGraw-Hill, New York).
COX, K.R. (ed.) [1978]
    Urbanization and Conflict in Market Societies, (Methuen, London).

DAHL, R.A. [1971]
   Polyarchy; Participation and Opposition, (Yale University Press, New Haven).
EVANS-PRITCHARD, E.E. [1937]
   The Nuer, (Clarendon Press, Oxford).
GOFFMAN, E. [1959]
   The Presentation of Self in Everyday Life, (Doubleday, New York).
HARVEY, D.W. [1973]
   Social Justice and the City, (Edward Arnold, London).
LEFEBVRE, H. [1968]
   La Droit a la Ville, (Anthropos, Paris).
POSPISIL, L.J. [1963]
   Kapauku Papuan Economy, (Yale University Publications in Anthropology, No. 54, New Haven).
RAWLS, J. [1971]
   A Theory of Justice, (Harvard University Press, Cambridge).
REMY, J. and L. VOYE [1974]
   La Ville et l'Urbanisation, (Duculot, Gembloux).
REMY, J. and L. VOYE [1980]
   Ville, Ordre, Violence, (PUF, Paris).
SANSOT, P. *et al.* [1978]
   L'Espace et son Double, (Le Champ Urbain, Paris).

Chapter 7
POLITICAL INTEGRATION AND DIVISION IN PLURAL SOCIETIES — PROBLEMS OF RECOGNITION, MEASUREMENT AND SALIENCE

J. Neville H. Douglas

To students of the political world the cleavages which divide human populations into politically distinct groups form an area of central concern. The most salient cleavages are formed in plural societies where long-standing antagonistic segmentation results in groups with separate political identities and aspirations (Lustick 1979). Cleavages in such cases lead not just to problems of effective administration within the political unit but to questions of legitimacy and right of existence. Clearly, problems of a different order arise in political units which contain plural societies.

Political integration and division can be considered as a consequence and as reflectors of cultural cleavages and political pluralism. Further study, however, shows that political integration can be viewed as both attribute and process (Merritt 1974). As attribute (the state of being integrated) it comprises a set of common cultural traits promoting a sense of community which provides the basis for accepted political institutions and laws which facilitate legitimate political activity and peaceful political change (Jacob and Teune 1964). As process (i.e. the act of integrating) it comprises a set of actions which binds a group together, creating cohesion and consensus and establishing political behaviour norms which become legitimised through time (Neuman 1976). While political division can be thought of as the converse of political integration, these phenomena should not be viewed as discrete elements and are best considered as opposite ends of a single continuum. Thus political units with plural societies containing diverse and conflicting processes will be located near the division end of the continuum while political units with homogeneous attributes and limited divisive behaviour will sit near the integration pole. In addition, the political unit can shift its position on the continuum as processes act to alter attributes and to integrate or divide.

The study of political integration and division has long been of interest to political geographers. Hartshorne (1950) argued the central importance of the raison d'etre of the state and emphasised the significance of centripetal (integrative) and centrifugal (divisive) forces (processes). Hartshorne's ideas were developed by Gottmann (1952) who noted the role of circulation and iconography within the political unit and by Jones (1954) who established clearly the relationship between cultural processes and the nature of the political unit. In subsequent work political geographers have paid attention to attributes such as language, religion and ethnic character (Alexander 1957; Pounds 1963; Muir 1975) and to the effects of process on state integration and division (Prescott 1968; Muir and Paddison 1981).

IDENTIFYING AND DESCRIBING POLITICAL INTEGRATION AND DIVISION

Integration and division are not immediately recognisable within the political unit, being dynamic phenomena which are dispositional in character. Like intelligence, with which Jacob and Teune (1964) draw interesting comparisons, integration and division have to be inferred from careful observation of selected indicators. As with intelligence testing, the question is: Which indicators among the attributes and processes form the best identifiers? Obvious indicators would seem to exist in the elements of political culture such as values and beliefs and in social and economic attributes. Such elements, individually or in combination, may indeed be used to identify and describe the presence of political integration or division yet at the same time these very elements can be causal factors influencing the levels of integration and division. Problems arise due to the tautology in identifying and describing integration and division in terms of the factors which control their existence. Cause and effect are so strongly intertwined that a fine line must be drawn between identifying element and causal factor.

In the search for the 'best identifier', community and the consensus connected with it provide the lead. Consensus, in political terms, relates to more than just a community but comes from and supports what Schmalenbach (1961) calls a communion. Communions are:

> . . . 'borne along by waves of emotion,
> reaching ecstatic heights of collective
> enthusiasm, rising from the depths of love
> or hate. . . They are bound together by the
> feeling actually experienced' (Schmalenbach
> 1961: 332).

Communion and consensus appear most readily in individual and group activities. The dynamism - the rise and fall - of communion and consensus is reflected in the varying intensity and change of attitudes. This suggests that political integration and division may best be identified by the study of attitudes. Other cultural attributes and behaviour processes may then be thought of as causal factors. As attitudes link back to fundamental human group values which act as prime motivators and forward to action choices in the real world, the distinction between cause and effect is tenuous, but is necessary to the better understanding of integration and division.

Attitudes are amenable to analysis as they are held towards individuals, groups, objects and situations. Perhaps the greatest difficulty arises when we ask: which individuals, groups, objects and situations engender the most salient attitudes as identifiers of political integration and division? The work of Sprout and Sprout (1966) and Boulding (1959) provide useful insights. Sprout and Sprout (1966: 501) suggest analysis should be concerned with:

(a) Attitudes toward the state, its authority and legitimacy
(b) Attitudes toward fellow group members of other political groups within the state
(c) Attitudes toward other states and their component political groups.

Boulding (1959) places similar emphasis upon group self images, attitudes toward the state and the nation and toward other groups and political outsiders.

As attitudes can change with new information flows in a dynamic environment, difficulty arises in analysis. To overcome such difficulty, social psychologists have developed alternative approaches such as measures in which inferences about attitudes are drawn from observed overt behaviour toward a chosen object or class of objects (Cook and Selltiz 1964). In studying political attitudes, measures might be developed from inferences drawn from voting behaviour, with the assumption that voting choice, certainly in plural societies, closely reflects underlying attitudes toward the state and its organisation. This method of assessment may arouse accusations of inconsistency when limiting of the identifying element to attitudes is immediately followed by description through study of behaviour. Yet it can be argued that voting behaviour stands in a separate class as a dispositional act which predates and conditions all subsequent political activity. As a reflector of political attitudes it builds in a spatial component which is seldom found in nationwide surveys of political attitudes (Taylor and Johnston 1979).

The importance of choice of identifying element can be illustrated by reference to Northern Ireland and its plural society. Religious affilation is the attribute most frequently used to identify the levels of political integration and division. The cleavage separating Protestants (62 per cent in 1980) and Catholics (38 per cent in 1980) is shown as sharp and clear and Northern Ireland is designated a two-segment plural society. Religious ascription however is taken not just as identifier and describing element but also as the causal variable. It is accepted implicitly that to be Catholic or Protestant is to have distinct cultural attributes and maintain different values, attitudes, and aspirations. A spatial element, sometimes incorporated as religious cleavage is shown to have a regional emphasis (Figure 7.1). As individuals seldom alter their religious affiliations the choice of this identifying element ensures a portrayal of political division and pluralism as intrenched and unchanging.

When attitudes are used as identifying element, a more subtle picture of cleavage, group allegiance and integration-division balance emerges. Attitudes toward the constitutional position of Northern Ireland in 1968 (Table 7.1) showed a clear majority (54 per cent) in favour, with 20 per cent against and a larger group of 26 per cent 'don't knows'. The limitations of the religious connotation, at least in 1968, are obvious. One out of every four approving the Constitution was Catholic and three out of every ten who disapproved of it were Protestant. The 'don't knows' were evenly spread between Catholics and Protestants. Similar evidence comes from attitudes toward the Border in 1968 (Table 7.2), one out of four of those who wished no change being Catholic. Attitudes toward power-sharing in 1978 (Table 7.3) again upset the traditional idea of two segments based on religious affiliation. The 64 per cent in favour of power-sharing was made up equally of Catholics and Protestants. Those opposed to power-sharing (19 per cent) were overwhelmingly Protestant.

It would seem that the element used to identify and describe integration and divisions conditions the perceived nature of the plural society very strongly. The over-dependence on the religious affiliation attribute for Northern Ireland creates a 'we' and 'they' image which even at times of greatest conflict is too exclusive. In consequence Northern Ireland society has become imprisoned in its own description.

When voting behaviour is used as the identifier of political integration and division another dimension is added to the description. As political parties in Northern Ireland vociferously establish ideological positions in relation to the Constitution, the Border, and power sharing, their levels of support reliably establish attitudes toward the fundamental political structures. Support for the Alliance Party of Northern Ireland represents support for the middle ground which rejects traditional political divisions (Table 7.4). Voting behaviour also shows variations in attitudes in time and space. Support for the Alliance Party is greatest at Local government level (Table 7.4) while the eastern area of Northern Ireland gives much stronger backing to the integrative policies of the Alliance Party than do the western and Border areas. This regional pattern parallels variations in attitudes among supporters of the Unionist Party. Support for pro-O'Neill Unionists (a more moderate and tentatively reformist faction) was found in the east while backing for the anti-O'Neill faction (staunchly traditionalist and wary of inter-party agreements) was located mostly in the western and southern peripheries (Figure 7.2).

50

Figure 7.1: Percentage distribution Roman Catholics, Northern Ireland, 1971.

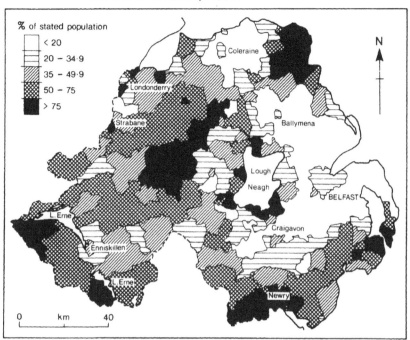

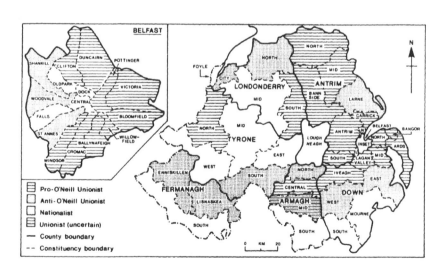

Figure 7.2: Northern Ireland (Stormont) Election 1969, Pattern of Unionist Voting

Table 7.1

Northern Ireland Attitudes Toward The Constitution

Question There has been a lot of controversy about the constitutional
position of Northern Ireland. On balance, do you approve or
disapprove of it?

Approve   54 per cent

Disapprove 20 per cent

Don't know 26 per cent

(Adapted from Rose 1971)

Table 7.2

Northern Ireland Attitudes Toward The Border 1968

Question (open-ended)
What change, if any would you like to see concerning the
Border?

Merge with Britain	6.7 per cent
No change	35.0 per cent
More co-operation across the Border	10.7 per cent
Abolish: alternative vague	28.3 per cent
Unite Ireland	8.0 per cent
Don't know	11.3 per cent

(Adapted from Rose 1971)

Table 7.3

Northern Ireland Attitudes Toward Power-sharing

A] May 1973
Question
How do you feel about power-sharing in the proposed Northern Ireland Executive? (Figures in percentages)

Strongly in favour	20.4
In favour	38.0
Do not mind	15.0
Willing to accept	9.3
Against	9.1
Strongly against	3.0
Don't know	5.2

(Source: Carrick James Market Research, Fortnight/Sunday Times Survey).

B] April 1974
Question
Do you approve or disapprove of power within the Executive being shared by parties representing the Protestant and Catholic communities in Northern Ireland?

Approve strongly	42.3
Just approve	30.9
Neither/don't know	9.6
Just disapprove	5.2
Disapprove strongly	12.0

(Source: NOP Market Research, Political Opinion in N. Ireland).

C] January 1978
Question
If Direct Rule was ended, would you be in favour of power- sharing (that is when representatives of all the political parties have a say in Cabinet Government) or would you be against it?

In favour	63.8
Against	19.1
Don't know	17.1

(Source: Opinion Research Centre, Northern Ireland Today, Ulster Television Survey).

Table 7.4

Support for the Alliance Party of Northern Ireland at Elections
1973-1981

		Votes	Percent of Total Vote
1973	Local Government	94,474	13.6
1977	Local Government	90,011	14.4
1981	Local Government	59,219	8.9
1974	N.I. Assembly	66,541	9.2
1975	N.I. Convention	64,657	9.8
1974(F)	General Election	22,660	3.1
1974(O)	General Election	35,955	5.1
1979	General Election		11.8

Support for Alliance Party of Northern Ireland in Local Government
Elections, 1977 and 1981. (Selected Districts).

	Percent of Total Vote in District	
	1977	1981
EAST		
Belfast (all)	18.6	13.2
Belfast South	33.8	27.7
North Down	38.5	25.2
Castlereagh	32.5	21.1
Carrickfergus	30.1	21.8
Newtownabbey	28.4	15.5
WEST/BORDER		
Londonderry	11.9	6.4
Newry and Mourne	8.3	3.6
Strabane	3.0	1.7
Fermanagh	1.9	1.6
Dungannon	1.7	nil

THE CAUSES OF POLITICAL INTEGRATION AND DIVISION

How can the formation of political integration and division be explained? In considering this question, it becomes evident that many factors contribute to the integration-division balance within the political unit. Attribute traits have been used most frequently as explanatory factors; few cleavages in attributes are regarded as a reflection of cultural homogeneity; this in turn is accepted as equating with political integration, whereas widespread cleavages and cultural heterogeneity are taken as certain reflectors of political division (Rae and Taylor 1970). Language, ethnicity and religion are usually accepted as the important attributes, together with the icons of group evolution such as knowledge and interpretations of historical events. In more recent studies of political integration, socio-economic attributes such as social class, occupation, employment, income and educational characteristics of the group have been considered significant (Coates, Johnston and Knox 1977; Cox 1979). The spatial expression of culture cleavage, as reflected in residential segregation of separate groups at regional, local and urban-rural scales, is also considered an important causal factor as physical distance has an effect upon the cognitive distance between groups (Jacob and Teune 1964).

It is noteworthy that the relationships of attribute cleavages have a significant influence on the strength of attributes as integrative and divisive factors (Figure 7.3). When cleavages regularly follow the same lines and repeatedly separate the population into the same attribute groups, the divisions in society become reinforced and intrenched. Conversely, when cleavages follow different lines and cross-cut within the population, the depth of division is reduced and potential for integrative behaviour is enhanced (Dahl 1976: 313).

Cultural processes (i.e. individual or group actions and operations) affect the levels of political integration and division. They introduce the dynamic element into the balance as the information flows and create or alter attitude patterns and intensities, encourage consensus or dissensus behaviour and thereby affect the position of the political unit on the integration-division continuum. The processes at work within the political unit are many and often interrelated and their influence on integration and division varies in time and space. Most processes have significant locational implications, for example processes which establish or remove hospitals, schools, homes, factories, universities and government offices. Processes thus create gainers and losers and can strengthen integration in one locality (that favoured by the decision) while simultaneously bringing about division in another (the neglected locality). Thus the processes of integration and division are complex, diffuse and multilinear, with many strands of varying significance. However, processes resulting from government decisions and authoritative allocation are of primary importance. Policies

Figure 7.3: Frequency Distribution of Attitude Intensity
Towards Power-Sharing in Northern Ireland

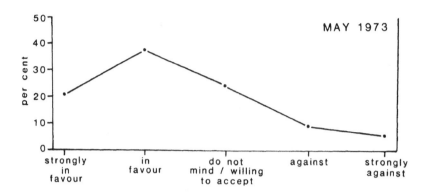

MAY 1973

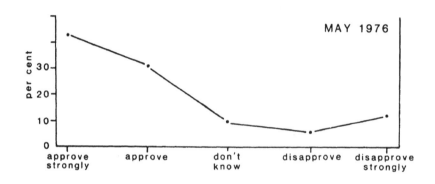

MAY 1976

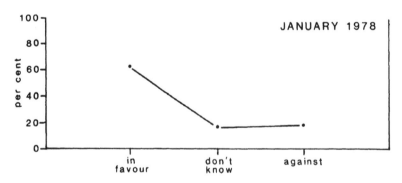

JANUARY 1978

(Source: K. Rose, I. McAllister, P. Mair 1978)

and resulting processes concerned with the reform of administration, local government and the electoral system – the redistribution of political power – form a fundamental category (Johnston 1979). Other public policies concerned with the allocation of scarce resources in the state economic and social sectors can have major effects. In the private sector, processes resulting from residential and educational decisions have a strong impact on political attitudes and so on levels of political integration and division.

Attribute and process factors are interdependent and work together to control the integration–division balance in the political unit. The nature and effect of processes are conditioned by the existing attributes of society while the attributes are the consequence of social processes at work over time. The relationship of attribute and process and the ways in which they combine to control political integration and division are shown in Figure 7.3.

Attempts to describe the range and relative importance of attributes and processes in the formation of political attitudes in Northern Ireland face problems. Difficulties arise from the established framework of description and explanation which focusses upon religious affiliation. Thus the data in Tables 7.1, 7.2 and 7.3 were set out, in their original form, by religious affiliation (Rose, McAllister and Mair 1978). Further consideration of attitudes by, for example, social class, education level, occupation, employed–unemployed or by urban–rural residence seldom takes place. The religious emphasis is continued further as analyses of social class, occupation, unemployment and education are carried out to show how the cleavages in such attributes relate to religious affiliation (Aunger 1975; Boyle 1976). Religion is taken as both the identifying and the explanatory variable, the implicit assumption being that religion and political attitudes are co-equal. As with the description, so the explanation of the Northern Ireland problem becomes trapped in the straitjacket of the single causal factor.

Osborne (1982) provides an alternative in his attempt to explain the Alliance vote in terms of socio-economic variables. The first preference vote in the 1973 Local District election was shown to have a strong positive association with adults possessing high educational qualifications, service sector employment and young adults and a negative association with poor quality housing, high unemployment, proportion of employed adults in agriculture and linear distance from Belfast. Studies such as this can help establish a wider explanatory framework and give greater insight into the causes of political attitudes.

The effect in Northern Ireland of cultural processes upon political attitudes has been widely considered and exemplified. Administrative policies and the consequent processes of local government and electoral area reform implemented by the Northern Ireland Government provide a good example. Numerous studies (Curran 1946; Gallagher 1956; Campaign for Social Justice in

Northern Ireland 1969) set out the reform policies since the first Local Government Act of 1922 as a power-intrenching, gerrymandering process perpetuating Unionist Government hegemony. Other works, including government papers and debates, set out the rationale behind the various policies. Whatever the reality underlying the layers of subjective argument, administrative reforms in Northern Ireland have had a fundamental effect upon intra- and inter-group attitudes (Douglas 1982). This is shown vividly by O'Donnell's (1977) study of Northern Irish stereotypes in which the minority nationalist group is shown as perceiving overwhelmingly the majority unionist group as simply 'power holders'.

Regional economic development policies and the associated processes of industrial attraction, factory location and job creation have also strongly affected group attitudes in Northern Ireland. A long-standing accusation is that government economic planning decisions have been politically motivated to favour the east (predominantly unionist) and disregard the west (predominantly nationalist) of Northern Ireland (Hoare 1981; 1982). Again, attitudes toward the Northern Ireland Government have been strongly affected, creating deeply suspicious views in the western regions of the country not just within the nationalist community but also, to an increasing degree throughout the 1960s, among unionists.

Other processes shown to influence attitude formation include separate school systems which increase social and cognitive distances (Buchanan 1982) and specific location policies which have placed Northern Ireland's new city at Craigavon and new university at Coleraine (Osborne and Singleton 1982). During 1981 the hunger strike policy of the Irish Republican Army prisoners created processes of emotion and conflict which polarised and intensified attitudes most strongly and sent Northern Ireland toward the division pole of the integration-division continuum.

ASSESSING POLITICAL INTEGRATION AND DIVISION

The acceptance of attitudes as identifiers of integration and division leads logically to the question of how the intensity of political integration and division and the salience of integrative and divisive factors can be measured and assessed. Additionally, the recognition that a wide range of attributes and processes affect attitudes leads to the question of salience, the strength with which any factor acts upon and influences an attitude. Intensity and salience are therefore key concepts and, in the absence of any general model of integration and division, it must be asked whether it is possible to develop quantitative measures of these concepts.

Several statistical measures concerned with levels of integration and division within the political unit have been developed. These are based largely upon the quantification of the number and extent of attribute cleavages within the unit. The cleavages fragment the population by setting members apart from one another and the extent of fragmentation suggests the importance of the attribute difference for levels of integration and division. For purposes of measurement, fragmentation denotes the number of elements or members of a population divided by a cleavage as a proportion of the total number of elements or population (Rae and Taylor 1970: 29). The indices, e.g. Fragmentation (Rae and Taylor 1970), Difference or Dissimilarity (Timms 1965), Differentiation (Muir 1975) and Segregation (Timms 1965) are all based upon such a form of measurement. Fragmentation is therefore a concept of wide utility as it can be applied to both attribute cleavages and attitude cleavages.

Each index establishes a method of assigning values to the degrees of fragmentation which lie between the two theoretical extremes of complete homogeneity and total fragmentation. Several questions present themselves.

First, is it possible to give valid theoretical interpretations of the values derived? The general argument related to the Index is that the higher the F (fragmentation) value, the greater the degree of division within the given unit. Thus, the Netherlands is shown to contain a much more divided society than Ireland or Norway (Table 7.5a). As Lijphart (1977) points out, however, in a study of consociational democracy in plural societies, the most difficult case for political integration and organisation is the two segment large minority - small majority plural society such as Northern Ireland. In such a case the F value will be lower than that of the more fragmented society such as the Netherlands in which several groups, all forming minorities, must of necessity, form coalitions and work together to form a majority and achieve power.

A second question concerns the classification of groups: how is it possible to establish the most politically significant set of groups within the political unit? In the example of religious affiliation, this question relates to the importance of the religious groupings chosen - are the chosen groups the most realistic to use in the derivation of the F value? In Table 7.5a, the F values are based on Roman Catholics, Protestants and non-Christian groups. While such a set may make good sense in most cases, it may not always be the optimum choice in terms of political integration and division, yet a common set of nominal groups is necessary if comparisons are to be made. It is also clear that by establishing a larger number of more specific groups, the value of F can be increased. So in the case of Northern Ireland, the F values can vary from 0.48 to 0.72 depending on religious groupings and from 0.64 to 0.79 based on different voting groups. The difficulty is that a priori assumptions about which groups are the most important must be made before using those same groups to derive a value for F which is then taken to reflect the salience of the cleavage.

Table 7.5a

Religious Fragmentation in Western States (1956)

(i)

State/Region	F Index (Religion)
Netherlands	.64
West Germany	.54
Canada	.54
Switzerland	.50
Australia	.50
New Zealand	.39
Austria	.18
Ireland	.11
Luxembourg	.05
Norway	.02
Northern Ireland (1971)	.48
Greater Belfast (1971)	.43

(Religious Groups: Roman Catholic Christian, Protestant Christian, non-Christian, including non-believers).

(ii)

Northern Ireland	.72

(Religious Groups: Roman Catholic, Presbyterian, Free Presbyterian, Church of Ireland, other Protestant, non-Christian).

(Source: D.W. Rae and M. Taylor 1970).

Table 7.5b

Voting Fragmentation (1977)

	F Index (Voting)
Northern Ireland	.64

(Voting Groups: Pro-partition, Anti-partition, Others)

Northern Ireland	.80

(Voting Groups: Loyalist Unionist, Official Unionist, Alliance Party, Social Democratic and Labour, Republican Clubs, Others).

(Source: S. Elliott and J. Smyth 1977).

60

Finally, the question of which areal unit scale to use cannot be overlooked. As Table 7.5a shows, the F value using the same nominal groups will vary with the unit of analysis - 0.48 for Northern Ireland as a single unit but 0.43 for Belfast as the most important subunit. It is clear that the relative salience of causal attributes remains elusive of effective quantitative measurement.

Studies of processes and the assessment of their integrative or divisive salience within the political unit have relied upon transaction flow analysis. This form of analysis follows from the Theory of Social Communication (Deutsch 1953) in which the most important factors integrating or dividing human groups are variations in the level and efficiency of social communication or interaction. The working hypothesis is that the volume of transactions reflects the level of integration existing within and between groups. The works of Mackay (1958), Savage and Deutsch (1960), Brams (1966) and Soja (1968) have developed the analysis and used it to draw conclusions about political integration and division. These representative works show the values and limitations of the transaction flow approach. Soja (1968: 50) establishes a relative acceptance index (RA) by comparing actual transactions with expected transactions between pairs of locations, the expected transactions being predicted by an indifference model. Soja was thus able to portray an informative pattern of linkages between the towns of Uganda, Kenya and Tanzania. However, the derivation of RA values between places prompts the problem of providing a valid theoretical explanation of their meaning. Soja argues that positive and salient levels of transaction exist between any pair of locations when actual linkages in both directions are 25 per cent greater than the predicted expected linkages. Unfortunately, no strong justification is given for the choice of a 25 per cent threshold, and the nature of the relationship between transactions, interaction and levels of political integration and division remains problematic. Any general hypothesis of the relationship probably has to be rejected in favour of a statement which recognises that the effect of transactions upon integration and division is controlled by the nature of the milieu in which the transactions take place. In environments where community conflict is strong, transactions between groups can, as Deutsch (1979) notes, hold greater potential for strengthening derogatory images and division than for political integration.

Another major problem of transaction flow analysis concerns the effect of distance on interaction between places. In the simple gravity model used by Mackay (1958) it is predicted that interaction and transactions will decrease at a rate proportional to the increase in geographical or physical distance, yet the recognition of the role of milieu points to greater importance of social and cognitive distance (Jacob and Teune 1964). This point has been made clear by Merritt (1964) in an historical study of Anglo-American interaction and by Boal (1969) in a study

61

of the Shankill-Falls divide in Belfast. To note the limitations
of general transaction flow measures is not, however, to reduce
the recognised importance of processes as factors influencing
political integration and division. Rather it is to emphasise
that general measures must be supplemented by detailed studies
of specific processes. In such studies, the growing use of
decision-making models within political geography should prove
valuable (Cox 1979; Muir and Paddison 1981). Such models can
provide the context in which the student of political integration
and division is able to isolate a decision, monitor the
implementation process and, most important, concentrate upon its
environmental consequences in terms of attitude formation.

Attitudes as identifiers of political integration and
division are held with varying intensity. Attitude intensity
is vital to understanding the nature of integration and division.
It has implications for action choices and future political
behaviour and controls the dispositons toward the intrenchment
of established political positions or toward attitude change and
political flexibility. It controls the outcomes from and the
possible success of policies concerned with encouraging consensus
and socially engineering integrative behaviour. It also influences
the work of the politician or the group leader seeking to maintain
support or group identity. In short, attitude intensity controls
group strength and the depth of political cleavages and its
analysis can best explain the position of the political unit on
the integration-division continuum (Warren and Jahoda 1976; Reich
and Adcock 1976).

However, as with attribute measurement, there are formidable
problems in assessing and measuring attitude intensity. One
problem concerns the difficulty of making clear the context in
which the attitude intensity develops. Forming an attitude
toward an object involves choice from among a set of alternatives
and the choice of one option means that others are rejected. The
intensity with which the attitude is held depends on the nature
of the alternatives. In Northern Ireland, support for power-
sharing has been gauged on a scale from 'strongly in favour'
through 'don't know' to 'strongly against' (Table 7.3). Among
the unionist community the location of attitudes on this scale
depends on whether the alternative proposed is 'good' (majority
rule regional government), 'satisficing' (continued direct rule)
or 'bad' (the reunification of Ireland). A 'good' alternative
is likely to reduce intensity of support while a 'bad' alternative
will intensify it. When alternative choices are practically
impossible and are perceived differently by individuals and
groups, the assessment of attitude intensity on a common and
comparable base becomes a major methodological obstacle (Rae and
Taylor 1970: 47).

The second problem follows from the nature of the individual's
intensity scales. Two individuals may each express an attitude
strongly against power-sharing, but it does not necessarily
follow that such common expression will have the same consequences

in terms of action choices, intrenchment or attitude change in future situations. Like integration and division, attitude intensity is stretched along a continuum and nominal groupings will involve a range of intensity and a range of consequences within each group. Aggregation of individual intensity scales therefore faces problems.

Despite these difficulties, it is barely possible to conceive of studies of political integration and division which do not take into account and attempt to assess the intensity with which individuals and groups hold to their attitudes and positions. The most commonly used approach is that which sets out a uni-dimensional intensity scale with intensity ranked from 'most favourable' to 'most unfavourable'. Attitudes are then located on the scale to produce a frequency distribution of attitude intensity within the given political unit. As Dahl (1956: Chapter 4) shows, consideration of such frequency distributions in relation to majority/minority problems can produce useful insights into the structural characteristics of political integration and division and their effects upon regime stability.

The frequency distributions of attitudes in Northern Ireland in 1973, 1976 and 1978 (Figure 7.4) show a considerable level of support for power-sharing with support being particularly intense in 1976. The distribution in 1973 displays a surprising amount of consensus and reveals a degree of integration not normally associated with Northern Ireland. However, each successive attitude survey also shows a perceptible growth in the ranks 'against' and particularly 'strongly against' power-sharing. This trend has probably continued since 1978 with consensus being steadily replaced by division and disagreement (Boal and Douglas 1982). Despite the limited data, the frequency distributions provide clear information on the structure of attitude intensities and encourage further questions on failure to implement power-sharing in Northern Ireland in the 1970s. Questions can also be asked about the distribution of attitude intensity within Northern Ireland as frequency distribution, perhaps reflecting the pattern shown in Figure 7.2, would establish stronger severe and symmetric disagreement with increased proximity to the Irish Border.

Despite these problems, careful survey and analysis using frequency distributions can lead to useful theoretical and practical considerations concerning political stability, degrees of political cleavage and the formulation of political rules and policies in varied attitude intensity contexts. The current problem in Northern Ireland is to develop political rules and a form of regional government to cope with a range of attitude intensity distribution, including a consensus pattern of some specific issues but, more frequently, a severe and symmetrical disagreement pattern on fundamental matters of constitutional position and political power.

Figure 7.4: Integration and Division Attributes and Processes

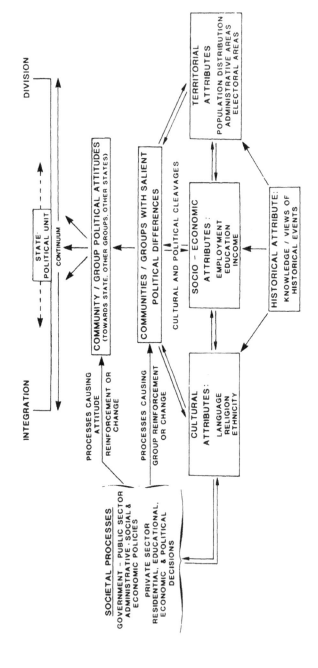

CONCLUSION

This wide-ranging consideration of political integration
and division shows that although much useful work has been carried
out since Hartshorne (1950) emphasised the importance of the
raison d'etre of the state, much remains to be done.

Quantitative measures of attribute differences and process
transactions provide a first step towards understanding the
salience of causal factors; yet such measures also serve to
highlight the complexity of integration and division. Despite
difficulties, the acceptance of attitudes as the identifying
element and the detailed analysis of attitude intensities may
possibly provide the most interesting and promising way to a
better understanding of the deeper nature of integration and
division. Nevertheless, the concepts remain elusive, with regular
shifts and variations in time and space obstructing effective
generalisation. Recognising this massive complexity and
interdependence of cause and effect leads to the realisation
that students of integration and division are attempting to
relate political consciousness and behaviour to the whole world
of political and social relations and structures. Such a
realisation may, in the end, lead students towards an ethnographic
approach (Burton 1978) to the study of integration and division
based on involvement, participation and observation at the local
scale.

REFERENCES

ALEXANDER, L.M. [1957]
    World Political Patterns, (Rand McNally, Chicago).
ALLIANCE PARTY OF NORTHERN IRELAND [1973]
    Local Government Elections Manifesto, (Belfast).
AUNGER, E.A. [1975]
    'Religion and occupational class in Northern Ireland',
    Economic and Social Review, 7, 1-17.
BOAL, F.W. [1969]
    'Territoriality on the Shankill-Falls Divide', Irish
    Geography, 6, 30-50.
BOAL, F.W. and J.N.H. DOUGLAS (eds.) [1982]
    Integration and Division: Geographical Perspectives on the
    Northern Ireland Problem, (Academic Press, London).
BOULDING, K.E. [1959]
    'National images and international systems', Journal of
    Conflict Resolution, 6, 120-31.
BOYLE, J.F. [1976]
    'Educational attainment, occupational achievement and
    religion in Northern Ireland', Economic and Social Review,
    8, 79-100.

BRAMS, S.J. [1966]
   'Transaction flows in the international system', American
   Political Science Review, 60, 880-97.
BUCHANAN, R.H. [1982]
   'The Planter and the Gael', in F.W. Boal and J.N.H. Douglas
   (eds.), Integration and Division, 49-74.
BURTON, F. [1978]
   The Politics of Legitimacy: Struggles in a Belfast Community,
   (Routledge and Kegan Paul, London).
CAMPAIGN FOR SOCIAL JUSTICE IN NORTHERN IRELAND [1969]
   The Plain Truth, (Dungannon).
COATES, B.E., R.J. JOHNSTON and P.L. KNOX [1977]
   Geography and Inequality, (Oxford University Press,
   Oxford)
COOK, S.W. and C. SELLTIZ [1964]
   'A multiple indicator approach to attitude measurement',
   Psychological Bulletin, 62, 36-55.
COX, K.R. [1979]
   Location and Public Problems, (Blackwell, Oxford).
CURRAN, F. [1946]
   Ireland's Fascist City, (Derry Journal, Londonderry).
DAHL, R.A. [1956]
   A Preface to Democratic Theory, (University of Chicago
   Press, Chicago).
DAHL, R.A. [1976]
   Pluralist Democracy in the United States, (Rand McNally,
   Chicago).
DEUTSCH, K.W. [1953]
   Nationalism and Social Communication, (M.I.T. Press,
   Cambridge).
DEUTSCH, K.W. [1979]
   Tides Among Nations, (Free Press, New York).
DOUGLAS, J.N.H. [1982]
   'Northern Ireland: spatial frameworks and community
   relations', in F.W. Boal and J.N.H. Douglas (eds.),
   Integration and Division, 75-104.
GALLAGHER, F. [1956]
   The Indivisible Island, (Gollancz, London).
GOTTMANN, J. [1952]
   'The political partitioning of our world: an attempt at
   analysis', World Politics, 4, 512-19.
HARTSHORNE, R. [1950]
   'The functional approach to political geography', Annals
   of the Association of American Geographers, 40, 95-130.
HOARE, A.G. [1982]
   'Problem region and regional problem' in F.W. Boal and
   J.N.H. Douglas (eds.), Integration and Division, 195-
   224.

HOARE, A.G. [1981]
    'Why they go where they go: the political imagery of
    industrial location', Transactions of the Institute of
    British Geographers, N.S. 6, 152-75.
JACOB, P.E. and H. TEUNE [1964]
    'The integrative process: guidelines for analysis of the
    bases of political community', in P.E. Jacob and J.R.
    Toscano (eds.), The Integration of Political Communities,
    (Lippincott, Philadelphia).
JOHNSTON, R.J. [1979]
    Political, Electoral and Spatial Systems, (Clarendon Press,
    Oxford).
JONES, S.B. [1954]
    'A unified field theory of political geography', Annals
    of the Association of American Geographers, 44, 111-23.
LIJPHART, A. [1977]
    Democracy in Plural Societies, (Yale University Press, New
    Haven).
LUSTICK, I. [1979]
    'Stability in deeply divided societies: consociationalism
    versus control', World Politics, 31, 325-44.
MACKAY, J.R. [1958]
    'The interactance hypothesis and boundaries in Canada',
    Canadian Geographer, 11, 1-8.
MERRITT, R.L. [1964]
    'Distance and interaction among political communities',
    General Systems, 9, 255-63.
MERRITT, R.L. [1974]
    'Locational aspects of political integration', in K.R.
    Cox, D.R. Reynolds and S. Rokkan (eds.), Locational
    Approaches to Power and Conflict, (Wiley, New York).
MUIR, R. [1975]
    Modern Political Geography, (Macmillan, London).
MUIR, R. and R. PADDISON [1981]
    Politics, Geography and Behaviour, (Methuen, London).
NEUMAN, S.G. [1976]
    Small States and Segmented Societies: National Political
    Integration in a Global Environment, (Praeger, New
    York).
O'DONNELL, E.E. [1977]
    Northern Irish Stereotypes, (College of Industrial Relations,
    Dublin).
OSBORNE, R.D. [1982]
    'Voting behaviour in Northern Ireland 1921-1977', in F.W.
    Boal and J.N.H. Douglas (eds.), Integration and Division,
    137-66.
OSBORNE, R.D. and D. SINGLETON [1982]
    'Political processes and behaviour', in F.W. Boal and
    J.N.H. Douglas (eds.), Integration and Division, 167-
    94.

POUNDS, N.J.G. [1963]
  Political Geography, (McGraw-Hill, New York).
PRESCOTT, J.R.V. [1968]
  The Geography of State Policies, (Hutchinson, London).
RAE, D.W. and M. TAYLOR [1970]
  The Analysis of Political Cleavages, (Yale University
  Press, New Haven).
REICH, B. and C. ADCOCK [1976]
  Values, Attitudes and Behaviour Change, (Methuen,
  London).
ROSE, R., I. McALLISTER and P. MAIR [1978]
  'Is there a concurring majority about Northern Ireland?',
  Studies in Public Policy, 22, (Centre for the Study of
  Public Policy, University of Strathclyde, Glasgow).
SAVAGE, R.I. and K.W. DEUTSCH [1960]
  'A statistical model of the gross analysis of transaction
  flows', Econometrica, 28, 551-72.
SCHMALENBACH, H. [1961]
  'The sociological category of communion', in T. Parsons
  et al. (eds.), Theories of Society, 1, 331-347.
SOJA. E.J. [1968]
  'Communications and territorial integration in East Africa:
  an introduction to transaction flow analysis', East Lakes
  Geographer, 4, 39-57.
SPROUT H. and M. SPROUT [1966]
  The Foundation of International Politics, (Van Nostrand,
  Princeton).
TAYLOR, P.J. and R.J. JOHNSTON [1979]
  Geography of Elections, (Penguin, Harmondsworth).
TIMMS, D. [1965]
  'Quantitative techniques in urban social geography', in
  R.J. Chorley and P. Haggett (eds.), Frontiers in Geographical
  Teaching, (Methuen, London).
WARREN, N. and M. JAHODA [1976]
  Attitudes, (Penguin, Harmondsworth).

Chapter 8
DILEMMAS OF PLURALISM IN THE UNITED STATES

Richard L. Morrill

INTRODUCTION

Throughout history, there has been a tension between forces of separatism and integration. Indeed, these are basic aspects of spatial behaviour and strong determinants of spatial organisation. Both have contributed to the richness of human society and the development of civilisation, and both have been at the root of severe conflict. The dilemma is how to reconcile or balance these forces in a society. Specifically for the United States, to what degree of diversity do benefits of enrichment outweigh possible costs of disunity and conflict? As with other basic dilemmas such as the balance between individual freedom and social responsibility, there can be no definitive answer. For some, the preservation of cultural differences may be worth the destruction of society, while for others, national unity may justify destroying cultural diversity. The intention here is only to develop some of the implications of a pluralist perspective.

The present era is one of renewed emphasis on cultural pluralism, or the preservation of differences. The present 'swing of the pendulum' toward the separatist end of the continuum is revealed not only through a sharpened racial and linguistic cultural identity, but also through the resurgence of neighbourhoods, the pursuit of local community interests, the revival of notions of autarchy, or local self-sufficiency and even the reaction against large-scale government business and advanced technology (Newman 1973). There are many reasons to account for why this trend should be so pronounced at this time. It is clearly an extension of the civil rights movement within the United States and decolonisation world-wide. While it is true that the civil rights movements in the 1950s and 1960s was integrationist in motive, it is not surprising that it shifted toward separatism, both because of white resistance and because of the emergence of stronger group leaders (Carmichael and Hamilton 1968). It is closely related to the recent emphasis on individual or group self expression. It is in part a reaction to perceived powerlessness in the midst of large cities and corporations. And, perhaps, in the face of the realisation of the homogenising power of modern communications and interdependence, many people may perceive the value of preserving some diversity before it is too late (Dahl 1978; Wiebe 1975). In any event, the purpose here is to examine how far and in what ways this shift has taken place, and to evaluate possible positive and negative effects (Baskin 1971; Isaacs 1975).

The structuralist and Marxist critiques suggest that social science serves to sustain societal norms (Peet 1977). Thus social scientists, as a group, appear not to have behaved independently, but to have shifted quickly with the spirit of the times. Although it appears from the literature that most were in the service of an integrationist philosophy in the 1950s and early 1960s, many appeared quick to embrace the values inherent in a separatist philosophy in the last dozen years.

PREHISTORY

While the intent here is to stress current issues in American society some wider background may be useful. The hundreds of thousands of years of human pre-history were the setting of much of this tension between separatism and integration, but we can only speculate as to its character. We do know that much of the basis for separatism today was laid then. The most obvious is the division of humanity into races of different colour – a product of long periods of evolution in varying physical environments. The beginnings of the differences in language systems and value system are also due to physical isolation of peoples in distinct environments over long periods. This differentiation of cultural traits may not have had to result in cultural conflict and tension between separatism and integration but it did, nevertheless.

Yet these were also millenia of human restlessness, millenia which long since saw intermixture, amalgamation and hybridisation of race and language over wide areas. Presumably much of the restlessness was forced on some groups by the inadequacy of their local environments and the resultant integration occurred as groups conquered one another or merely occupied the same richer environments.

It is probably safe to say that most human groups are conservative, in that they will try to defend what they have and what they have become. It is also probably safe to say that, in general, the smaller the group the less likely was significant technical innovation and that the rise of civilisation awaited the successful coordination of larger numbers in wider areas. But for whatever reasons – land, slaves, prestige – repeatedly some groups have successfully conquered others and attempted to govern territories that did encompass cultural differences.

ROMAN IMPERIALISM AND EUROPEAN DEVELOPMENT

In the historic period, it will be useful to examine briefly the Roman Empire. This was a very practical empire which brought together highly diverse peoples under an administrative umbrella, essentially for political and economic purposes. On the one hand, the unity and vitality of the empire was sufficient to lead to some important commonality over a wide area, such as in Roman law, administrative structure, engineering and agricultural technology, and eventually Romance languages and Christianity. Consider only the value of the imposition of a standard width for wagons and carriages, which carried over into standardisation of railway systems! It cannot be denied that this integration of peoples was stupendously beneficial and creative, and that this forced amalgamation made possible whole levels of human achievements which would not otherwise have occurred. On the other hand, Rome was sufficiently tolerant to respect and even adopt localisms from around the empire, and this was ultimately expressed, for example, in the growth of many Romance languages rather than the one Latin. On the other hand there were limits to Roman tolerance of diversity as the destruction of Jerusalem attests.

In the end, the power of the centre was not sufficient to control the forces of separatism. For example, local languages were gradually permitted for local reporting and administration. Peripheral groups no longer had reason to identify positively with Rome, and the region returned to an era in which, except for the Christian Church, localism once more prevailed.

The post-Roman development of Europe illustrates similarly the tension between separatism and integration and why it is not really possible to assert that one or the other is necessarily good or bad. On the one hand the localism of the feudal period led to much of the exciting cultural diversity that is Europe. On the other hand, if nationhood based on ethnicity (e.g., language or race) had been inversely pursued, then many European nations of today would not exist, nor would their complex and rich cultures have arisen. For example, it cannot be denied that Great Britain has made a stupendous contribution, through its language, literature, technology, democratic institutions, to civilisation. Yet, if in the 12th century, nations had been formed on the basis of ethnicity there should have been three or four Celtic states, a Danish state, and a Norman one, and an England might never have developed. And despite its abuses, if Great Britain had not conquered and maintained a vast empire for more than two centuries, its contribution would undoubtedly have been but a fraction of what it has been. France, Italy, and Germany, too, are not at all purist survivals, but new cultures and languages growing out of the merging of formerly separate groups.

In light of European history, one must wonder seriously about the seeming permanence with which boundaries of dozens of new, small nations are viewed. The point of this review is to remind us that the social, political and geographical structure of Europe is the legacy of centuries of resolving conflicts between these forces of separatism and integration, not to argue that one or the other was 'better'.

THE UNITED STATES

The United States is in some respects similar to the Roman Empire. In 1800 the U.S., while politically independent, was clearly a British cultural outpost. The U.S. adopted a fairly generous policy, even if the motivation was to occupy the land and provide depth in space and numbers. Even before 1850 this resulted in significant numbers of German and Scandinavian as well as British immigrants. In effect, the national policy was to grant land and economic opportunities, even freedom of religion, in exchange for the transfer of political allegiance to a new nation and for the gradual acceptance of the English language and unifying aspects of British culture. The underlying mood – and a brilliant strategy really – was that the nation could amalgamate, incorporate, universalise many cultural elements of the immigrant groups, and even permit groups the partial maintenance of cultural differences, as long as these were secondary to joining the mainstream culture. This constituted a rather constrained form of pluralism. Although perhaps now in disrepute, the melting pot really did work for most groups, throughout most of American history. In the most critical respects, embracing a common language, law and economy (Handlin 1959; Ward 1971), it is undeniable that the very existence and survival of the nation depended on the success and acceptance of this admitted cultural imperialism. In fact virtually all groups of immigrants accepted this quid pro quo. It is perhaps significant that most immigrants came of their own free will and voluntarily changed allegiance and modified their culture, while the Black population was forced into slavery, and the Indian (Native American) population was conquered. The practical limits to cultural diversity were set mainly in the marketplace and community but also by government and courts (Dahl 1976). For example, the admission of Utah as a state was conditioned on the abolition of bigamy. The U.S. Civil War was certainly a test of the acceptable limits of a regional versus a national culture. Would the degree of pluralism embraced today have respected the secession of the South?

Pluralist sentiment in the 1970's moved far toward relinquishing this primary tool of assimilation. Maintenance of the original language became of primary importance and the acceptance of English became secondary as the case of American Indians and Spanish-speaking groups illustrate. The argument was that language is the key cultural right and that the United States, like Canada, is sufficiently strong to survive greater internal variety. Many nations can and do live with more than one language.

A more important question is whether it is wise from the point of view of the group's own long-term self interest. The United States is likely to absorb millions more migrants from Mexico. To encourage their separation and isolation by emphasising their cultural and linguistic uniqueness may serve to condemn them to a much longer period of economic exploitation and perceived second-class status. Bilingual instruction, from my reading of history, should be a means toward integration of linguistic minorities into the general economy and culture, not a shield against it.

Language was in fact an important and emotional issue to many immigrant groups. They accepted the necessity to learn English but also needed the security of their native language and institutions. These dual purposes were served by the fact that most groups were spatially clustered in ethnic ghettoes and by the development of such ethnic institutions as newspapers, cultural centres and language schools. These institutions persist as does some degree of geographic segregation, but most people view these as sources of cultural heritage appreciation rather than as primary or alternate cultures as such.

In general, the greater the perceived difference in characteristics of a group, the more the dominant group will fear it and discriminate against it. Thus in American experience, discrimination was least severe against new white Protestant groups, the next least severe against white Catholics, then white Jews, then not-so-white Mexicans and Indians and Asians, and most severe against blacks (Kenyon 1976).

Discrimination against blacks has only begun to relax after centuries of severity. Despite changes in law the perception that they were unwelcome in mainstream white institutions led many blacks for a while to a strategy of emphasising cultural differences and pursuing power through group consolidation and identification, and this has led effectively to successful court recognition of a _right_ of political and economic representation because of colour.

Success in registering blacks resulted in dramatic increases in black political representation at all levels, but only if they remained spatially segregated. This created an extremely difficult dilemma for the black community between the benefits of group self-esteem and power through remaining separate and electing more officials, and the benefits to many individuals, and probably to the group in the longer run of pushing to integrate.

The same legal approach has been pursued with reasonable success by _women_, as a _class_, both of which have a constitutional basis in the 14th Amendment. This solution appeared necessary because of systematic long-term discrimination in these cases against different cultures, but of groups obviously a part of the mainstream culture but not permitted a fair share in its opportunities.

This formal recognition of group rights also extends, in voting and redistricting, to the American Indians and the population of Hispanic origin. Unfortunately these precedents have led to a wider separatism, and to the general proposition that every group, for whatever purpose, deserves representation and special protection merely because of its existence.

With respect to voting and redistricting the courts have accepted no further claims. Thus in United Jewish Organization of Williamsburg vs. Carey, 1977, the Supreme Court held that redistricting which split the Jewish community did not violate the constitutional provision for equal protection. Similarly claims of unfair division of social or economic or political groupings have been dismissed.

Societies may, for humanitarian reasons, recognise the special needs of some groups - for example, the elderly and the handicapped, or mothers of independent children, or the mentally ill, which are unable to participate in the society without some help-- but there may be a real risk of carrying group identity too far. The most important interests and needs of the majority of people are common - for example, health, employment, housing and education. Yet the effect of cultural pluralism and geographic localism is to seek separate institutions: separate schools, separate clinics and hospitals, separate housing structures for categories of people: e.g., hospitals for American Indians, housing for the elderly, schoolrooms for Chicanos, which at best fosters the idea that health, education, housing or employment means something inherently different to different groups, which is nonsense, or at worst, leads to inevitable variation in quality and, as a consequence, in outcomes. The history of separate institutions - Indian schools, public health, veterans and county hospitals for the poor, urban renewal housing for the poor, decades of separate schools for blacks - has been a dismal one (Wolpert and Wolpert 1975).

The problem is that while separate institutions give some recognition to the group, it also tends psychologically to stamp them as inferior, a constant reminder that they cannot compete or participate equally. Studies of the Chicago hospital system revealed the immense financial, health and social costs of having separate facilities for the poor and for Blacks (De Vise 1973).

The current alliance of conservatives who defend inequality, and of many liberals, who espouse pluralism, against the racial integration of schools is from an historic standpoint a particularly fascinating example of the separationist - integrationist dilemma (Itzkoff 1969; Ravitch 1976; Smith and Ley 1974). School integration, through court ordered bussing has not led to a quick or dramatic increase in black proficiency, and did accelerate white flight to suburbs and private schools. Yet the willing abandonment of integration risks substituting more severe long term distrust and conflict for short term disruption and disappointment. Admittedly the lesson from school integration is that one institution alone cannot overcome centuries of separation and discrimination, including employment discrimination and housing segregation. The rational minority person seeking to advance his or her status and well being would reasonably reject separate facilities as inherently discriminatory and demand full participation in mainstream institutions and opportunities.

This analysis does not suggest the suppression of valued cultural elements, including language. But that Americans are a part of an integrated economy implies that the highest priority should be paid to achieving successful participation in the economy through acceptance of the lingua franca and the general norms of society. Without this a group appears doomed to an inferior status that can be in part self-imposed. The recent Vietnamese and Chilean refugee immigrants appreciate this very well, and appear to be able to adopt a new culture, without abandoning valued elements of their original culture.

The case of the American Indian is perhaps the most sensitive and complex case, at least in the U.S. Pacific Northwest. The history of the treatment of the Indian has been grim, including the virtual expropriation of land, the introduction of disease, and systematic economic, social and political discrimination over the years. For both the defeated Indian nations and the dominant white culture, the separatist-integrationist dilemma has been most difficult, and attitudes and policies have varied over the years. Despite periods of suppression, language and other aspects of culture have survived. This is in large measure a consequence of the separation of Indians onto reservations, a kind of recognition of nationhood, but of an inferior and proscribed sort. Yet given the nature of the integrated national economy, was it not rather the crowning cruelty to separate Indians physically onto remote reservations, so that they could not integrate into the society, and yet could not begin to provide the economic basis for a meaningful independent existence, either in their traditional or in the mainstream American ways of life? Only recently have improvements in transportation and communication, and the energy demands of the wider society penetrated the reservations, forcing both sides to deal with the issue of separatism and integration anew. But it is a delusion to pretend, on either side, that the present culture of any but the most remote Indians is really distinct or that they are not inescapably a part of the general American economy. It is no longer realistic to attempt to recreate a lost state of cultural purity or to achieve political-economic independence, but here, too, to seek that 'almost but not quite' assimilation that characterised most immigrant groups.

PLURALISM AND GEOGRAPHIC SCALE

The dilemma of pluralism and the costs and benefits of separatism and integration operate at all spatial and organisational scales from within the household to the world as a whole (Cox 1979). In general the effects of a pluralist ascendancy is to decentralise power and proliferate specialised institutions. Thus at the world scale the United Nations is weak since international relations operate under the pluralist assumption of political and cultural sovereignty of states. The process is carried so far as to justify, for the majority of nations, not only strict control of movements, but also of information. The human costs, the risks of misunderstanding and war resulting from intense and jealous nationalism are stupendous, but the record of integrating forces, for example the multi-national corporation or the U.S. and Western Europe in television and films is not very comforting, either.

Since states are the critical organising units for society and for the use of land and resources, it is perhaps the level for which pluralism holds the greatest risks, for which the dilemma of separatism and integration is most constantly in mind. However, the local scale is where the pluralist diversity of a nation actually appears on the landscape – not so much within a local area or jurisdiction but across them. Thus there is a sorting out on the landscape of cultural diversity – a spatial specialisation that reduces the frequency of contact and conflict, but likewise fails to resolve differences or prevent severe conflict. Where local diversity is too great, conflict occurs, as can be seen in the record of U.S. racial disorders.

At all scales the tension between separatism and integration is also expressed geographically, through the relationship of core and periphery. In general the dominant core of a country or region or city attempts to impose its ways on the periphery. In other words, it is integrationist in motive and action. Peripheral areas try to protect their diversity and autonomy, hence the proliferation of separate suburbs around the city but the attempts of the core city to create metropolitan structures (Buchanan, 1977). Economically and politically, the more integrated the national or regional economy and communications net, the more the initial advantage and momentum of the core can be maintained.

INTEGRATION AND PLURALISM

The exciting thing about the American experiment is that enough amalgamation of diverse cultures occurred to enrich all of them and to create a powerful, creative new culture, obviously far more than the sum of those parts (Glazer and Moynihan, 1975). It is not just a transplanted British culture, but a new culture that has made vast contributions to the rest of the world. This could not have occurred if the U.S. had opted for separate educational systems, separate areas of settlement, separate health systems, separate unions and firms for people of different

76

national origin or language, and each religion, for example, for Protestant, Jewish and Roman Catholic German immigrants (Stein and Hill 1977). That is the structure of institutions which has proven so disastrous to Lebanon, perhaps Northern Ireland, and threatens, in a less extreme way, the integrity of Belgium, Canada, Nigeria, and many other nations.

There is a peculiar irony, a measure of the intellectual confusion of the times, that it was the liberals who strongly embraced the new cultural pluralism, who espoused neighbourhood power and the interests of the narrower against the wider good (Mazziotti, 1974). In fact, it often served the interests of conservatism, a reaction to the admitted truth that integration and sweeping national policies had not ushered in a golden age, that it was not really possible or worthwhile to overcome inequality. Liberals should ponder, in advocating the maintenance of separate Indian nations or Spanish language zones, that this tends to condemn such groups to classic exploitation as a cheap labour reserve. Indeed it well serves the interests of capital to help maintain an aspect of cultural diversity, which tends to weaken the bargaining power of solidarity of labour. The divisions of society are real - by colour, region, class but the glorification of differences cannot offer much beyond symbolic comfort to the less successful and ultimately risks more conflict. Progress occurs from the confrontation and resolution of ideas and problems and peoples. It does not occur from a passive coexistence.

The above discussion has stressed the risks of a pluralist and separatist philosophy, both on the integrity of a nation and on the wider progress of humanity - that it is fundamentally a conservative stance, out of a fear of change. On the positive side, acceptance and encouragement of cultural diversity has the potential of preserving diversity in custom and traits that is not only fascinating and enriching, but useful when shared with other cultures (Greeley 1976; Logue 1976; Triandis 1976). The suppression of other cultures by conquerors who imposed an integrationist culture has also been common in history and the results destructive - for example, the Ottoman role in the Balkans, or the extreme post-Moorish Catholicism in Spain, the sorry record of religious persecution and wars in Europe and elsewhere, and most hideous, the Nazi genocide of Jews and Gypsies. And the relative tolerance of diversity in the United States contrasts to the tendency toward cultural homogeneity in most other countries - for example, in restrictions on acceptable names for children, in prevention of the entry of useful foreign words into the language, in required support of the state churches, or in the rigid censorship of outside influences. Pluralism presents risks, but it does protect diversity, foster tolerance of others, decentralises power and provides more group and individual freedom.

So the answer to the dilemma is not at all simple. It is easy to be a critic of America. But the U. S. has been more successful than most in striking a balance, imposing a cultural homogeneity in a few basic dimensions such as the use of English, the dominance of public non-sectarian education, belief in an integrated, national economy, with no restriction of the internal movement of peoples and goods while permitting a strong measure of localism, via a federal political structure, extreme tolerance of religious diversity, including a parallel structure of languages and religious schools and publications, and a stupendous array of organisations and societies devoted to as many special interests. The warnings are not against the latter, but not to forget the former - that another can afford the luxury of cultural diversity only to the extent that it also maintains and respects those elements of culture that conserve the whole society. Perhaps the best measure of this is that groups and areas and their representatives are able to recognise the existence of legitimate wider interests than themselves.

REFERENCES

BASKIN, D. [1971]
American Pluralist Democracy, (Van Nostrand, Princeton)
BUCHANAN, K. [1977]
'Economic growth and cultural liquidation', in R. Peet (ed.), Radical Geography, (Maaroufa Press, Chicago), 125-42.
CARMICHAEL, S. and C. HAMILTON [1968]
Black Power; The Politics of Liberation, (Random House, New York).
COX, K.R. [1979]
Location and Public Problems, (Maaroufa Press, Chicago).
DAHL, R. [1976]
Democracy in the United States, (Rand McNally, Chicago).
DAHL, R. [1978]
'Pluralism revisited', Comparative Politics, 10, 191-212.
DE VISE, P. [1973]
Misplaced and Misused Hospitals and Doctors, Resource Paper #22, (Association of American Geographers, Washington D.C.).
GLAZER, N. and D. MOYNIHAN [1975]
Ethnicity; Theory and Experience, (Harvard University Press, Cambridge).
GREELEY, A. [1976]
'The ethnic miracle', Public Interest, 45, 20-36.

HANDLIN, O. [1959]
   The Newcomers, (Harvard University Press, Cambridge).
ISAACS, H. [1975]
   Idols of the Tribe, (Harper and Row, New York).
ITZKOFF, S. [1969]
   Cultural Pluralism and American Education, (International
   Textbook Company, Scranton, PA).
KENYON, J. [1976]
   'Patterns of residential integration in a bicultural Western
   city', Professional Geographer, 28, 40-44.
LOGUE, E. [1976]
   'The idea of America is choice', in Qualities of Life,
   (Lexington), 1-52.
MAZZIOTTI, D.F. [1974]
   'Underlying assumptions of advocacy planning; pluralism
   and reform', Journal of the American Instititute of
   Planners, 40, 38-48.
NEWMAN, W. [1973]
   American Pluralism, (Harper and Row, New York).
PEET, R. (ed.), [1977]
   Radical Geography, (Maaroufa Press, Chicago).
RAVITCH, D. [1976]
   'Integration, segregation and pluralism', American Scholar,
   45, 206-17.
SMITH, C. and D. LEY [1974]
   'Ethnic pluralism, competition and discretion over primary
   education in Montreal', in Community Participation and the
   Spatial Order of the City, (University of British Columbia,
   Vancouver).
STEIN, H.F. and R.F. HILL [1977]
   'Limits to ethnicity', American Scholar, 46, 185-89.
TRIANDIS, H. [1976]
   'Future of pluralism', Journal of Social Issues, 32,
   179-208.
WARD, D. [1971]
   Cities and Immigrants, (Oxford University Press,
   Oxford).
WIEBE, R. [1975]
   The Segmented Society, (Pergamon, Oxford).
WOLPERT, E. and J. WOLPERT [1975]
   'From asylum to ghetto', in S. Gale and E. Moore (eds.),
   The Manipulated City, (Maaroufa Press, Chicago), 329-
   38.

Chapter 9
THE CHANGING SITUATION OF MAJORITY AND MINORITY AND ITS SPATIAL EXPRESSION — THE CASE OF THE ARAB MINORITY IN ISRAEL
Arnon Soffer

INTRODUCTION

This chapter examines the importance of majority, minority and equal status feelings and the geographic expression of these feelings in pluralistic states. Most work on majority/minority relationships has taken the form of description, demographic analysis and political consequence and, more often, the overt and covert struggles resulting from these complex situations. Here, it is shown that in plural states in which the definition of what constitutes a majority and minority varies from group to group and from area to area, the resulting fears and images held by each group towards others lead to actions and reactions to ensure group existence, all of which have clear spatial expression. To provide a concrete example, the relationships between Jews and Arabs in Northern Israel are examined.

MAJORITY-MINORITY RELATIONSHIPS IN PLURALISTIC STATES

Over 90 per cent of the world's independent states have some form of pluralistic structure.

Boal (1983), in discussing multiple minorities, presents the case of Northern Ireland in which both Protestants and Catholics simultaneously constitute minority and majority. The Catholics are part of the majority when they describe the territory as including the Republic of Ireland and the Protestants are the majority in Northern Ireland, a minority in the island of Ireland and part of the Protestant or British majority in the United Kingdom. Boal has analysed these situations in different regions and towns of Northern Ireland and shows how the majority/minority relationships vary according to the different territories. In spite of the discussion which deals with all the majority/minority viewpoints in Northern Ireland in the past and present and in spite of the stress placed on the fears accompanying each group in relation to its neighbour, he does not deal with the spatial results of these fears.

In Cyprus, the ethnic structure resembles that of Northern Ireland. The Turks are a clear minority on the island - 19 per cent in 1969 (Drury 1972), but when considered along with the Turkish mainland, they constitute a majority in relation to the Greeks on the island and part of the majority even when compared with all Greeks. The Greeks, on the other hand, constitute a large majority on the island and they managed to retain this majority until the Turkish invasion in 1974 when a Turkish controlled territory was created by force in the northern part of Cyprus.

The continuing claims of the Turkish population for additional rights and representation in the government of the island sprang from the support that they received from Turkey due to the majority feeling that was evident. In contrast, on the island itself, in a day-to-day context, they formed a very marked minority which lagged behind the Greek population on socio-economic criteria (Drury 1972).

However, Greek antagonism toward the Turkish population sprang from their majority status and its attendant power on the island, leading to the continuing claims for union with Greece (Enosis). The Turkish invasion and its bitter consequences are the result of a mistaken Greek assessment based on majority/minority status on the island alone rather than in the whole region. The Turkish invasion followed a continued period of violence, initially with many clashes among peasants with later violence in the towns, culminating in an all-out war with several cease-fires (Middle East Economic Digest 1974-1975; Time 1974; Cypri 1975).

Even more complex situations are provided by Lebanon and Yugoslavia. In Lebanon, the Sunni Moslems form the second largest group after the Shi'ites according to a 1982 estimate. Yet, they are part of the Sunni Arab majority in the states which surround Lebanon and this feeling of majority has been expressed in political aspirations towards union with Syria (Bannerman 1980; Smith 1974). The Maronite Christians who were formerly the largest group, now occupy a lower position and find themselves as a majority only in the heartland of Mount Lebanon. This led, in the past, to the actual creation of an independent Lebanese state and has brought considerable political power to this group (Smith 1974). It is generally recognised that the Shi'ite Moslems today form the largest group in the Lebanese mosaic. Formerly, they were a marginal minority group within Lebanon, a position similar to that which the Shi'ites occupy in the Middle East as a whole. However, they form a considerable majority in Southern Lebanon and events of the past 25 years have brought about the creation of a separate Shi'ite political party and militia (Zamir 1981). Even the Palestinians in Lebanon suffer the split image of majority-minority - a minority within Lebanon but with considerable support throughout the Middle East, they succeeded in controlling large areas of Southwestern Lebanon from 1976 to 1982.

Although the ethnic structure of Lebanon has been widely studied (Beaumont, Blake and Wagstaff 1976; Fisher 1972; 1978; Hartman 1979), little attention has been paid to such questions as mutual proximity of the various groups, economic dependence and interdependence and their role in the disintegration of the state and the influence of the relative isolation on the level of activity within the national system.

A similar, if even more complex situation, exists in Yugoslavia (Alexander 1963; Carlson 1958; Weigert 1957; Borowiec 1977). As a result of the linguistic, religious and ethnic divisions within the state, a federal state with six autonomous republics (Serbia, Croatia, Slovenia, Bosnia-Herzegovina, Macedonia and Montenegro) was created, giving partial expression to an even more complex ethnic structure.

Minority groups, such as the Albanians of the Kosovo Region and Serbia bordering on Albania, constitute part of a large majority in their own region and when combined with Albanians on the other side of the border. On the other hand, the Albanians as a whole are a minority within the Balkan states. A similar situation exists for the Macedonians, Roumanians, Bulgarians, Germans, Turks and Hungarians within Yugoslavia's borders who are not only minorities in the regions in which they live but were formerly supported by external nation states.

The three large groups in the Yugoslav core have a different set of problems. Croats and Slovenes fear Serbian hegemony and strongly support a federal structure whereas the Serbs press for a greater degree of centralisation. Many questions exist as to the nature of group interrelationships in Yugoslavia and the level of representation at national level of each group.

Many more examples of majority-minority dilemmas can be cited with perhaps the Sunni Arab - Shi'ite Arab - Shi'ite Iranian - Sunni Kurd complex of Iraq and Iran outstanding in this respect.

MINORITY-MAJORITY RELATIONSHIPS

Muir and Paddison (1981: 165) note that

> 'Conflicts whose origins stem from the
> multi-ethnic composition of the state are
> among the most intractable [with] which
> governments have to contend. . . . their
> severity can be great enough to threaten the
> territorial integrity of the state.'

They use the model constructed by Rabushka and Shepsle in order to show patterns of interethnic behaviour within states which strongly reflect their numerical balance. Rabushka and Shepsle (1972: 88) present four situations:

1) Dominant majority
2) Balanced competition
3) Dominant minority
4) Fragmentation

Although Rabushka and Shepsle examined situations which lead to an undermining of democratic regimes in plural societies, their model is also suitable for examining the whole range of relationships between the majority and minority in any country.

The first situation is one in which the majority suppresses the minority and the minority, in order to survive, takes undemocratic action, as in Northern Ireland, Sri Lanka, Cyprus, Rwanda and Zanzibar.

The second situation is where the majority of the population and the minority are almost equal from a demographic point of view. Every attempt to create harmony between them usually fails and a solution lies in the direction of federal structure, as in the examples provided by Malaysia and Belgium.

A third situation is where the minority controls the majority. South Africa and Rhodesia provide examples for this situation, which is insufferable and invites suppression and rebellion.

The fourth situation is typical of new African states such as Nigeria, Sudan and Zaire and also of Lebanon, where there is a varied network of tribe-nations.

The examples outlined above show the explosive potential in the many varieties of plural societies which exist. Often national tragedies are involved and, in some cases, conflict is ongoing. These cases verify the Rabushka-Shepsle model with regard to the confrontation which can be expected in plural society and justify the comment by Muir and Paddison on the pressing need for governments to come to grips with these problems.

Whereas the Rabushka-Shepsle model relates only to events within a state, there is a need to expand it to cover areas outside the state, such as two states or a state and a neighbouring region where a minority in one state regards itself as part of a majority in another separate, but closely related territory.

Two different situations exist when a minority within a state is concentrated in a small area in which it can be considered a majority and when the minority is distributed randomly over the state area. Different behaviours will result from these two cases.

These examples indicate that a fifth situation, in addition to the four outlined in the model, exists. In this fifth situation, Situation 1 (majority), Situation 2 (equal status) and Situation 3 (minority) are exhibited simultaneously, as a result of variable definitions of territory only.

ARABS AND JEWS IN ISRAEL

Israel is a pluralistic society, consisting essentially of two main cultures, Jewish and Arab, each tending to live apart from the other and different in religion, social and cultural values and in national expectations (Smooha 1976: 407-08).

Those Arabs on the Israeli side of the 'Green Line' (the pre-1967 border between Israel and its neighbours) are a small (17 per cent) minority within Israel. However, the Arabs of Galilee (almost half of Israel's Arab population) form half the population of Israel's Northern District and they form 71 per cent of the population in the hill areas. A similar situation can be observed amongst the Arabs of the 'Little Triangle' in Central Israel (Brawer, Chapter 13, this volume) and in Jerusalem (see Figure 9.1).

The Arabs in Israel can be shown as a majority, as a co-equal entity or a minority of the population in relation to Jews, depending on the scale examined. This numerical exercise can be run in the opposite direction and the Arabs can be described as a substantial (35 per cent) minority within the boundaries of British Mandate Palestine (Eretz Israel) or at an even broader scale, examining relations between Jews and Arabs throughout the Middle East, the Jews become a minority, representing a mere 3 per cent of the total population (Table 9.1).

In this study, we deal with group relations in larger geographic regions rather than those in towns and neighbourhoods or villages, although studies at such scales have been carried out in Israel, especially in mixed towns (Cohen 1973; Kipnis and Schnell 1978; Waterman 1980; Ben-Artzi 1980).

Table 9.1 presents a state of affairs that differs from that which prevailed in Palestine prior to 1948. The situation then was more positive from the Arab viewpoint as they formed a majority throughout the country with the exception of Jerusalem. Thus, images and memories have developed among each group which cannot be ignored when studying different vistas on this complex and interacting system.

Different forecasts for various regions within Eretz- Israel can raise hopes on the one side and fears on the other which may influence future actions and reactions. From an analysis of the population forecast for Eretz-Israel in 1995, the major expected change is in the Arab population of Northern Israel which will by then have attained majority status, and a further strengthening of the Arab demographic position in the hill areas of the North (Table 9.2).

JEWISH-ARAB PERCEPTIONS

How do the Jewish and Arab inhabitants of Galilee regard themselves and one another from the point of view of national identity?

Figure 9.1: Israel - Arab Concentrations

In recent research on the independent identity of Arab Israelis (and in some cases, Jewish attitudes towards them), the essential question centres around two attitudes:
1)   Do they feel themselves to be Israelis or do they feel themselves a minority within the State of Israel? and
2) Do they feel themselves to be Palestinians? (thus attesting to a feeling of belonging to a larger group equal to or larger than the Jewish group within the region).

The results indicate confusion and difficulty in defining the identity of Israel's Arabs (Tessler 1977: Hoffman and Beit Hallahmi 1976; Smooha 1980b: 58-60). Smooha also asked about identification with the term 'Arab'. Of the total Arabs interviewed, 10.3 per cent answered that they feel themselves Arabs, thus belonging to the majority group in the widest possible region; 45 per cent defined themselves as Israeli Arabs and 35.1 per cent as Palestinian Arabs. As opposed to this, 71.5 per cent of the Jews interviewed defined the same population as Israeli Arabs and 13.2 per cent simply as Arabs (Smooha 1980b: 60). Smooha (1976) also discussed the majority/minority relations between Jews and Arabs from the sociological point of view. However, no attempt has been made to link majority/minority status to events in the landscape nor to examine the effect of geography on the creation of such feelings.

ISRAELI ARABS AS AN ISRAELI MINORITY: ARAB EXPRESSION

The feeling of the minority status of the Arabs is expressed in political advertisements, in the Knesset (Israel's parliament) at demonstrations and in private conversation in which charges of discrimination against the Arab minority, damage to their rights and displacement from their lands are aired. Smooha (1976; 1980a; 1980b: 92-109) has examined the feelings of grievance and his research shows that almost all the Arabs in Israel expressed a feeling of having been treated unfairly.

Several Arabs have written about this feeling (Hareven 1981).

> 'We see our national rights also in the
> removal of discrimination and national
> suppression. . . .in the differences in
> grants made to local authorities and in
> regards education. . . .There is discrimination
> in the Citizenship Act. . .(Tuma 1977).

The claim is essentially justified. Arab settlements are discriminated against when compared with the level of services in Jewish settlements. There are many reasons for this, but the main one is that of government aid to the Arab sector. As a result, there is a clear expression of this discrimination in the Israeli landscape.

Table 9.1

The Majority/Minority Status of Jews and Arabs in Selected
Areas c. 1980

REGION	JEWS		ARABS		TOTAL	JEWS FORM
	'000s	%	'000s	%	'000s	
Middle East*	3,300	3	108,000	97	111,500	m
Palestine	3,300	65	1,800	35	5,100	M
Israel	3,300	83	640	17	3,940	M
N. District	317	52	296	48	613	E
Galilee Mts.	105	29	260	71	365	m
Hadera and						
C. District	820	86	130	14	650	M
Little Triangle	10	8	117	92	127	m
Metro Jerusalem	300	60	200	40	500	M
Jerusalem	292	72	115	28	407	M

(Sources: Statistical Yearbook of Israel 1981, Statesman's
Yearbook 1981-1982)

*   The Arab countries only: Egypt, Syria, Lebanon, Jordan
    Saudi Arabia, Iraq, Kuwait, Bahrein, U.A.E., Qatar,
    N. Yemen, S. Yemen, Oman, Sudan, Libya.

    M = Majority m = Minority E = Equal status

Table 9.2

Population Projection For Selected Regions in Eretz-Yisrael
1995

REGION	JEWS per cent	ARABS per cent	JEWS FORM
Israel	79	21	Majority
N. District	48	52	Minority
Galilee Mts.	25	75	Minority

As Rabushka and Shepsle indicate in their model, the Arab minority has also taken several democratic steps aimed at preserving its rights and survival; demonstrations, legal protest, court cases against land appropriation or confiscation, disbelief in the intentions of the government (Rekhess 1977: 11-17) and voting for the Communist Party which supports their rights (Rekhess 1977; Standel 1972; Landau 1971; Smooha 1976: 413). Nondemocratic action has also been taken, including illegal building inside and outside the villages on disputed or government-owned lands, illegal political organisation (such as el-Ard), sabotage and espionage.

JEWISH EXPRESSION

The status of Israeli Arabs has been expressed in all levels of Government (Rekhess 1977; Smooha 1976; 1980a) through the established institutions (Porat 1981: 204-08), in the newspapers, in literature and in daily contact.

The perception of the Arabs as a weak minority has allowed the introduction of several laws aimed at regulating the problem of land ownership in Israel, such as the Acquisition of Lands Law, 1953; the Cultivation of Waste Lands Ordinance, 1948; the Absentee Property Law, 1950 (Porat 1981: 205). Without these laws, it would have been impossible to establish new towns such as Upper Nazareth or Carmiel.

Since the Arabs are a minority, it has been possible since 1948 to replan the Galilee radically and establish Jewish settlements there. In some instances, the removal of Arab settlements was involved; in others, this involved the establishment of permanent Bedouin settlements, often with substantial government benefits.

As economic power was in the hands of the Jews, this created a situation of economic dependence on the part of the Arabs, establishing dominance (Lustick 1980; Smooha 1980a: 19). The spatial consequences of this are expressed in commuting, in the lack of industrial establishments in the Arab sector and in a transportation network leading from the Arab villages to the major employment centres.

EQUALITY BETWEEN JEWS AND ARABS IN NORTHERN ISRAEL

Since 1948, the Arabs of Northern Israel have comprised half of the population of the region. In 1948, the population of the Northern District consisted of 37 per cent Jews; in 1961, the proportion had risen to 57.6 per cent and in 1980, it had fallen to 51.6 per cent (Statistical Yearbook 1981: 32).

There exists a situation of co-equality in economic matters and in whole sectors one can point to total interdependence, as, for example, in commerce and industry and in areas of contact between the groups, such as in Nazareth and Nazareth and in Carmiel and the Beit Kerem Valley. Another example is provided by Acre, Nahariya and Haifa, which act as industrial and commercial centres for both sectors in the North (Bar-Gal and Soffer 1981).

There is Jewish dependence on Arab manpower, principally in agriculture, building and personal services and Arab dependence on Jews in industry, finance, commerce and public services (Soffer, in press). As the Rabushka and Shepsle model suggests, the feeling of demographic equality and dissatisfaction with the unequal distribution of resources form the background to the demands for greater equality.

> . . . 'to fight so that the government's ministries guarantee equal treatment to all local authorities in Israel and so that they act towards the Arab local authorities as they do with the Jewish.'

Such were the claims of the heads of Arab Local Councils in 1974 (Rekhess 1977: 17).

The foundation of a Committee for the Protection of Arab Lands and the subsequent Committee of Heads of Arab Local Authorities is a result of the feeling of equality which did not exist in the 1970s. It is no accident that the organisation began in Northern Israel where the feeling of equality exists, and not in Jerusalem, amongst the Bedouin of the Negev or in Central Israel.

The Jewish feeling of equality with the Arabs in the North is connected to the constant fear that the delicate demographic balance in the area will be upset and this acts as the source for the strategy of Jewish settlement in the whole area (Al HaMishmar 1976).

THE ARABS AS A MAJORITY: ARAB EXPRESSION

The Arabs of Israel are part of the whole Arab world which surrounds Israel. They are connected to this world through family, cultural, religious, social and political ties, and through radio and television.

Smooha (1980b: 45) found that 60 per cent of the Arab population in Israel watch television or listen to radio programmes broadcast from the neighbouring countries. They are connected with the Arab countries through meetings overseas and physically since 1967 through the 'Open Bridges' over the River Jordan established by Israel.

Since 1967, when the 'Green Line' ceased to exist, contacts between the Arabs of Galilee, the Gaza Strip and the West Bank grew, as did personal contact with Arabs throughout the Middle

East. The contact with the Arabs in the Occupied Territories has had far-reaching results which have not yet been studied in sufficient depth. The conclusions of the National Committee for the Protection of Arab Lands after 'Land Day' in 1976 stated that 'the solidarity expressed by the residents of the conquered areas in their declaration of a general strike and demonstrations. . .is natural solidarity for the sons of one nation split by the Imperialist-Zionist connection. . .' (The Black Book 1970: 26). This is what leads to a feeling of being part of a decisive majority (Situation 1 in the Rabushka-Shepsle model). This feeling is expressed in the poem by Tawfiq Zayyad, Member of Knesset and Mayor of Nazareth, 'We are the majority here'.

> 'The vain robbers say that we are a minority who will be in exile one day. . .but. . .are we a minority? No! No! A million times No! <u>For we are the majority here</u>. . . and in this enchanted strip of land, here, in proud Galilee and in Nazareth, <u>we have a homeland</u> . . . and the nation which has been uprooted over there . . . <u>we will bring it back</u> and make it into a new paradise.'
> (Yinnun 1981: 238-239).

This feeling is expressed in Arab propaganda which equates Israel with South Africa and concludes from this that Israel should become a democratic-secular state according to the formula set down by the P.L.O. whose practical meaning is a return to the natural former state of an Arab majority and a dominated Jewish minority, and the elimination of the Jewish Zionist entity of Israel (Maoz 1978: 546). The feeling of belonging to the whole Arab world is reinforced by the reality of their being a decisive Arab majority within the territory of the upland areas of Galilee.

The expression of this reality can be found in the demands for separation and irredentism and even calling for the destruction of Israel by some extremist groups such as the 'Sons of the Village'.

It is also expressed in the elections to the Knesset in 1977 and 1981. In the heart of Galilee and in the Arab urban centres, the vote for the New Communist Party was very strong, accounting in some instances for up to 80 per cent of the voters. The party is the most radical of the legal parties in terms of demands for improving the lot of the Arab sector.

The power of this party, especially in the large towns in the centre of Galilee, points to a connection between the feeling of being a majority there and in their rejection of the Zionist State of Israel. (The Druze and Bedouin and the Arabs in the smaller towns and villages gave a smaller vote to the Communists).

One expression of the feeling of power was the general strike in March 1976 (Land Day) which acted as a sort of civil insurrection against the land policy in Israel.

> '. . .to turn this day into a 'Land Day' in Israel when the Arab masses will raise their voice in a demand to end the evil policy which has turned into a threat to our future in this land.' (Rekhess 1977: 57).

In contrast to other rural societies throughout the world, a characteristic of the Arab sector in Israel is the low rate of migration from the villages to the large towns despite the fact that this population has been commuting to the cities daily for 25 years.

Some workers are known to commute 120 kilometres every day in each direction. Several researchers have studied this phenomenon (Bar-Gal and Soffer 1976) and, amongst the factors which explain it, they mention the political conflict and the socio-cultural gap between the Jewish population which is predominantly urban and the Arab population which is rural. Other factors are the close family ties within the traditional Arab society and the generally short distances between the villages and the centres of employment. They also mention the high cost of apartments in the large cities as opposed to the housing situation in the villages.

One might suppose that the feeling that every (isolated) Arab feels in the Galilee, of being part of a majority, actually increases his security there and acts as a complement to the factors noted in regard to his non-migration from the village to the city. The lack of cityward migration, in addition to high natural growth rates, has led to another important consequence. The Arab villages have spread outwards and in many instances form continuous built-up areas which give rise, at least in their external appearance, to metropolitan areas, as in Nazareth and the Little Triangle (Bar-Gal and Soffer 1981). These metropolitan areas contain several thousand Arabs who form a considerable force by themselves. The fact that Christian Arabs have a tendency to emigrate from Israel altogether and the fact that the Moslems do not raises yet another point, namely that the Christians have little feeling of being in a majority in Israel or in the region as a whole and so do not see a future for their existence here while the Moslems are more bound to the place being part of the overall majority.

JEWISH EXPRESSION

The Jews themselves do not deny the feeling that the Arabs are a majority in the area. Layish (1981: 243) writes:

Figure 9.2: Jewish and Arab Settlements in Galilee

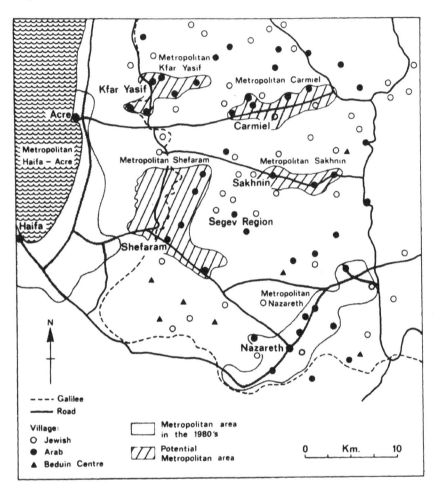

Figure 9.3:   Jerusalem Metropolitan Area

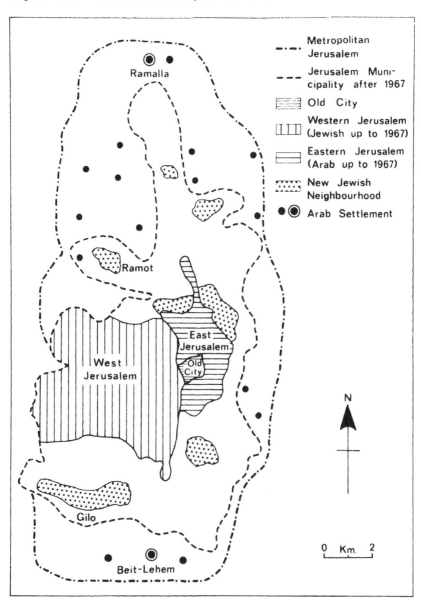

>'This recent transformation of the Arabs
from a majority during the Mandatory period
to a minority in Israel has not yet become
rooted in the consciousness of the two
sectors. The differences between majority
and minority on the socio-economic- cultural
level is easily translated at times of crisis
into national-political terms.'

Porat (1981: 198) also relates to this point and states:

>'The fear of general Arab enmity and the
anxiety over a uniform negative stand taken
by Arab states regarding Israel has exacerbated
the less than friendly relations between
Jews and Arabs inside Israel'.

Smooha (1980b: 67-73 and Lustick (1980) point out that Jewish
supervision over the Arab sector in Israel has its source in
fear of the majority.

This feeling of the existence of an Arab majority in the
Galilee and in the whole region is the moving force for all the
development policy and Jewish settlement in Eretz Israel,
especially in the Galilee.

On a national level, the idea which guided the execution
of the 'Allon Plan' (settlement of Jews in the Jordan Valley
after 1967) was the need to cut off the Arabs of the West Bank
from Arabs on the other side of the Jordan in addition to the
military and strategic reasons. Similarly, the Rafiah development
was established to be a barrier between the Gaza Strip and Egypt.
For this reason the Jews created a 'security belt' around the
Lebanese and Syrian borders where they did not permit Arabs to
reside.

The fear that a contiguity of Arab settlement in the defined
area will lead to demands for separation or even to irredentism,
is what guides the Jewish settlement strategy there since 1948
and has even been given names such as the 'Judaisation of Galilee'
or 'Populating Galilee'.

The U.N. resolution concerning the division of Palestine
in 1947 contributes more than a little to this fear of autonomy
or irredentism, since the Galilee was attached to the proposed
Arab state. Muir and Paddison (1981: 160) discuss the existence
of danger to a state which leaves peripheral areas in the hands
of minorities and the need to guarantee the sovereignty of these
areas by entering them.

The clearest examples of this policy of entering Arab areas
in Galilee is the establishment of Upper Nazareth and Ma'alot
in the 1950s in the heart of 'purely' Arab areas. In the 1960s,
Carmiel was established in an Arab-populated area in the Beit
Kerem Valley. Jewish settlement efforts reached a peak in the
1970s with the establishment of the Segev settlement bloc as a

Jewish obstacle to Arab blocs which were developing in Sakhnin-Arabeh in the east and Tamra-Shfar'am in the west. In the second stage, many industrial villages and 'outlook settlements' were established close to Arab settlements and to thicken settlement along the Lebanese border (Soffer 1981) (see Figure 9.2).

The hesitation about enforcing Israeli law on the Arab population of Galilee is yet another example of the fear of the Arab majority in the area. This is marked by Arab squatters on State Lands in various parts of Galilee with little done to prevent it. A similar situation exists with regard to illegal and unauthorised building within and outside the villages on private agricultural land and even on state-owned lands (Bar-Gal and Soffer 1976; Soffer, in press).

In Jerusalem, we can observe a repeat performance of events in Galilee. The fear of an Arab majority in the city and its surrounds leading to a danger of pressures to redivide the city is the factor which has motivated government actions to judaise Jerusalem and to thicken it through a dense system of neighbourhoods, each of which is approximately the size of one of Israel's development towns (Figure 9.3).

A similar policy for similar reasons has developed in recent years in Judaea and Samaria (Rowley, this volume) in which a large number of settlements has been located in a strategic alignment in order if not to bring about a Jewish majority then to prevent the separation of the region from Israel.

SUMMARY

It has been shown that in Israel, a situation exists which was not taken into account by Rabushka and Shepsle. In this situation, a population group feels itself as majority, co- equal partner and minority at one and the same time. This feeling is also felt by the Jews toward to the Arabs.

This complicated structure is expressed in each of the two sectors in declarations, behaviour and deeds. If, according to the model presented, each of these situations of inter-group demographic relations leads to bitterness and the weakening of relations between the groups, then the three situations existing simultaneously represent a triple time-bomb.

Israel is not the only state in which groups within a plural society perceive themselves as minority and majority at the same time. This is the situation which Boal has termed 'multiple minorities', as this situation is true for countries such as Ireland, Lebanon, Cyprus, Yugoslavia and others, all of which have undergone violent conflicts, in which the level of violence tends to increase in proportion to the strength of parties with vested interests over the border.

The first stage in the conflict is the stage in which the groups with a feeling of relative strength attempt to claim more rights and resources than those which they already possess, thus leading to a situation of unquiet, mutual fears and lack of confidence.

The second stage leads the group to take the initiative and it has the most marked spatial expression. Settlement efforts occur in sensitive areas in order to prevent irredentism or to increase the level of interconnections between the periphery and the core.

The next stage is that of confrontation. This usually develops from local incidents eventually covering the whole country in a civil war and even entangling outside forces.

In addition, this paper would claim that it is necessary to examine yet another view on relations between groups in pluralistic societies, a view which has been inadequately examined until now. This is the spatial view and it is not only an expression of confrontation but also serves to catalyse that confrontation.

The geographic literature deals largely with the first stage with descriptions of ethnic and religious systems, and with the third stage, as the historical development of boundaries, the history of confrontation and the political solution of conflicts. However, much less work has been done on the second stage; scholars have not faced the dilemma of changing majority/minority relationships and the results of this dualism. Likewise, they have not stressed the dual role of the spatial expression in group conflict, serving as both the instigator of initiatives and as a catalyst which strengthens or weakens confrontation.

REFERENCES

ALEXANDER, L.M. [1963]
    World Political Patterns, (Rand McNally, Chicago).
AL HA-MISHMAR
    September 7, 1976 (Tel-Aviv)
BANNERMAN, M.G. [1980]
    'Republic of Lebanon', in D.E. Long and R. Bernard (eds.),
    The Government and Politics of the Middle East and North
    Africa, (Westview Press, Boulder), 207-30.
BAR-GAL, Y. and A. SOFFER [1976]
    Changes in Minority Villages in Israel, Horizons in
    Geography, 2, (in Hebrew).
BAR-GAL, Y. and A. SOFFER [1981]
    Geographical Changes in the Traditional Arab Villages in
    Northern Israel, (Centre for Middle Eastern and Islamic
    Studies, University of Durham, Durham)
BEAUMONT, P., G.H. BLAKE and J.M. WAGSTAFF [1976]
    The Middle East, (Wiley, London).

BEN-ARTZI, Y. [1980]
Residential Patterns and Intra-Urban Migration of Arabs
in Haifa, (Occasional Papers on the Middle East, (New
Series), No.1, University of Haifa, Haifa), (in Hebrew).
BOAL, F.W. [1983]
'Multiple minorities', in L.M. Whirter and K. Trew (eds.),
The Northern Ireland Conflict: Myth and Reality: Social
and Political Perspectives, (G.W. and A. Hesketh, Ormskirk
and California).
BOROWIEC, A. [1977]
Yugoslavia After Tito, (Praeger, New York).
CARLSON, L. [1956]
Geography and World Politics, (Prentice-Hall, Englewood
Cliffs).
COHEN, E. [1973]
Integration versus Separation in the Planning of Mixed
Arab-Jewish Cities in Israel, (Jerusalem), (in Hebrew).
CYPRI, J. [1975]
Background Notes on Cyprus, (Nicosia).
DAVAR
June 7, 1976 (Tel-Aviv)
DRURY, M.P. [1972]
'Cyprus: Ethnic dualism', in J.I. Clarke and W.B. Fisher
(eds.), Populations of the Middle East and North Africa,
(University of London Press, London), 161-81.
FISHER, W.B. [1972]
'Lebanon: an ecumenical refuge', in J.I. Clarke and W.B.
Fisher (eds.), Populations, 143-60.
FISHER, W.B. [1978]
The Middle East, Sixth edition, (Methuen, London).
HAREVEN, A. [1981]
One Out of Every Six Israelis, (Van Leer Foundation,
Jerusalem), (in Hebrew).
HARTMAN, K.P. [1979]
Tubinger Atlas des Vorderen Orients, (Reichert Verlag,
Wiesbaden).
HOFMAN, J.E. and B. BEIT HALAHMI [1978]
'The Palestinian identity and Israel's Arabs', in G. Ben-Dor
(ed.), The Palestinians and the Middle East Conflict,
(Turtledove Publishing, Ramat Gan), 215-28.
KIPNIS, B. and I. SCHNELL [1978]
'Changes in the distribution of Arabs in mixed Jewish-
Arab cities in Israel', Economic Geography, 54, 168-88.
LANDAU, J.M. [1971]
The Arabs in Israel, (Oxford University Press, London).
LAYISH, A. [1981]
'Trends after the Six-Day War', in A. Layish (ed.), The
Arabs of Israel: Continuity and Change, (Magnes Press,
Jerusalem), 241-47, (in Hebrew).

LUSTICK, I. [1980]
   Arabs in the Jewish State: Israel's Control of a National
   Minority, (University of Texas Press, Austin).
MA'ARIV
   November 14 1978, January 1 1979, February 21 1979 (Tel-
   Aviv).
MAOZ, M.
   'New attitudes of the P.L.O. regarding Palestine and
   Israel', in G. Ben-Dor (ed.), The Palestinians, 545-54.
MIDDLE EAST ECONOMIC DIGEST
   23 August 1974 - 25 April 1975
MINISTRY OF INTERIOR, MINISTRY OF FINANCE [1972]
   A Programme for the Geographic Spread of an Israeli
   Population of Five Million, (Jerusalem), (in Hebrew).
MUIR, R. and R. PADDISON [1981]
   Politics, Geography and Behaviour, (Methuen, London).
NATIONAL COMMITTEE FOR PROTECTION OF ARAB LANDS IN ISRAEL
[1976]
   The Black Book on 'Land Day', 30 March 1976, (Haifa)
NATIONAL PROGRESSIVE MOVEMENT
   18 January 1979
PORAT, Y. [1981]
   'The opinion of the secular Jew', in A. Hareven (ed.),
   One Out of Every Six, 198-210, (in Hebrew).
RABUSHKA, A. and K.A. SHEPSLE [1972]
   Politics in Plural Societies, (Charles Merrill,
   Columbus).
REKHESS, E. [1977]
   Arabs in Israel and Land Expropriations in the Galilee:
   Background, Events and Implications, (Shiloah Institute
   for Middle Eastern and African Studies, No. 53, Tel- Aviv),
   (in Hebrew).
SIKRON, M. [1983]
   'Changes in the population of the North', in N. Kliot, A.
   Soffer and A. Shmueli (eds.), The Lands of Galilee, (Haifa
   and Tel-Aviv), (in Hebrew).
SINGLETON, F.B. [1970]
   Yugoslavia, (Queen Anne Press, London).
SMITH, H.H. [1974]
   Area Handbook of Lebanon, (The American University,
   Washington, D.C.).
SMOOHA, S. [1976]
   'Arabs and Jews in Israel: Minority-majority relations',
   Megamot, 22, 397-423, (in Hebrew).
SMOOHA, S. [1980a]
   'Existing and alternative policy towards the Arabs in
   Israel', Megamot, 26, 30-52, (in Hebrew).
SMOOHA, S. [1980b]
   The Orientation and Politicisation of the Arab Minority
   in Israel, (Occasional Papers on the Middle East, New
   Series, No.2, University of Haifa, Haifa).

SOFFER, A. [1981]
    'Jewish settlement in Galilee 1948-1980', Israel Land and
    Nature, 10-16.
SOFFER, A. [forthcoming]
    'Geographical aspects of changes within the Arab community
    of Northern Israel', Middle Eastern Studies.
STATISTICAL ABSTRACT OF ISRAEL [1981]
    (Central Bureau of Statistics, Jerusalem).
STENDAL, O. [1972]
    The Minorities in Israel, (Jerusalem).
TESSLER, M. [1977]
    'Israel's Arabs and the Palestinian problem', Middle East
    Journal, 31, 313-29.
TIME
    'Cyprus', 29 July 1974
WATERMAN, S. [1980]
    'Alternative images in an Israeli town', Geoforum, 11,
    277-87.
WEIGERT, H.W. et al. [1957]
    Principles of Political Geography, (Appleton Century
    Croft).
WORKER'S COVENANT
    12 January 1979
YINNON, A. [1981]
    'Tawfiq Zayyad: We are the majority here', in A. Layish
    (ed.), The Arabs of Israel, 213-40.
ZAMIR, M. [1981]
    'The ethnic distribution as the root of the civil war in
    Lebanon', Monthly Survey, 6/23, 3-14, (in Hebrew).

Chapter 10
COMMUNAL CONFLICT IN JERUSALEM — THE SPREAD OF ULTRA-ORTHODOX NEIGHBOURHOODS

Yosseph Shilhav

The purpose of this paper is to understand the unique causes for the segregation of the ultra-Orthodox population of Jerusalem and to study the ways and means by which it exerts its influence on the urban system.

SEGREGATION AND ULTRA-ORTHODOX JEWS IN JERUSALEM

A tendency towards segregation of the ultra-Orthodox population of Jerusalem has been observed in recent years (Hershkovitz 1976).

A segregative spatial structure of population in urban areas is well known and has been studied over the years (Timms 1971; Herbert and Johnston 1978) and by many scholars. Erickson (1954) presented a general model of segregation in the urban population and described a dynamic process of change in the map of distinct areas in the city by way of invasion and succession. Processes of invasion of a given population into a residential area in which it was previously unrepresented were examined for blacks, a population known to accelerate segregation processes (Morrill 1965). Indeed, the racial factor as a cause for segregation is very strong, even stronger than economic parameters (Raymond 1970). This view has been supported by Rose (1971) who studied ethnic causes of segregation. He also stated that the ethnic causes are the most acute since they carry physical expression. Black residential areas, according to Rose, create a subsystem in the general urban structure.

Other causes of segregation are as a result of the actions of socio-economic variables and social stratification (Rossi 1955) and cultural factors, such as religion (Meinig 1965).

Segregation tends to create socially homgeneous residential areas in a city. The coordination of members of a particular group in order to further its particular interests with regard to the territorial search for residential areas is one of the main topics of study in urban political geography.

Ultra-Orthodox Jews

The population under study in this paper is a minority group in the Israeli population and as such tends to gather in distinct areas. Like all concentrations of special groups, it is motivated by a positive incentive - a wish to create a 'cultural dominance' in a given territory (Downs), in order to facilitate the socialisation processes of the younger generation, and by a negative incentive - a wish to prevent assimilation and social and cultural influences that are foreign to the character of the particular group (Wilson 1967; 1972). These factors, which are two sides of one coin, are universal and applicable wherever the phenomenon of voluntary minority segregation in urban areas is encountered. In this respect the ultra-Orthodox Jewry of Jerusalem does not differ from world-wide norms.

A closer look at the question of the geographical concentration of these ultra-Orthodox Jewish communities reveals, however, that there is a unique explanation for this spatial behaviour which is not usually applicable to other groups and by which other specific processes which accompany the territorial concentration and expansion of this community can be understood.

The segregational process of the ultra-Orthodox Jewish population in Jerusalem as well as in other Israeli towns, poses several problems of definition and identification. Although we are dealing with a case of geographical segregation based on religion, extra care must be taken as we are not dealing with different religions but with population groups that are all Jewish, but which differ from one another in their way of life. These differences stem from the various practical applications of 'being a Jew' as perceived by each of the groups. The wide variety of groups involved can be narrowed into two categories: those that observe all the religious commandments and those who do not. The ultra- Orthodox Jews represent the most extreme among those who practise the totality of religious commandments. Thus, we can define the ultra-Orthodox Jews as those whose way of life and culture are fully formed on the basis of Jewish law (the halacha in its most orthodox interpretation and on the basis of Jewish traditions as crystallised within the realm of Ashkenazi culture, mostly in Eastern Europe, during recent generations.

The segregation of this population in Jerusalem presents a high level of intolerance, associated with violence, towards any appearance of a secular way of life. This population will not tolerate any deviant living within the boundaries of its territory.

In order to understand the spatial behaviour of an Orthodox Jew, one must bear in mind that the orthodox interpretation of Judaism differs from the concept of religion as encountered in the Western world. It differs in two basic principles. First, Judaism is not just a religion involving a certain conception of the world but a religion of practical commandments which

101

prescribe the daily behaviour of the Jew. Secondly, Judaism is not restricted to matters that deal with the individual and his God but also involves matters which are of communal interest. Therefore, each member of the community is held responsible for the observance of the commandments by the other members as well (Babylonian Talmud, Sanhedrin, 27b). A simplified and uncompromising application of these principles, especially the second, does not permit the Orthodox Jew to live close to another Jew who does not observe religious laws.

As an illustration of this point, imagine an apartment house in which the residents own their apartments and which has an elevator. The elevator is part of the joint property; operation of the elevator on the Sabbath is prohibited according to the Orthodox interpretation of Jewish law. In an ultra-Orthodox interpretation, an Orthodox Jew living in the building is not only prevented from using the elevator on the Sabbath but, as part-owner, he must see to it that the other neighbours will not operate it. This may cause conflict between the Orthodox resident and his neighbours. If the Orthodox Jew would prefer to spare himself such conflict, he will avoid living in a building alongside non- observant Jews. Herein lies the crux of ultra-Orthodox residential segregation.

On a different scale, streets and neighbourhoods form a space common to all the population of a settlement. The ultra-Orthodox Jew considers himself responsible for the way of life practised in the common domain and therefore will not accept any behaviour which does not conform to Jewish law. He is left with the choice of trying to dictate the rules of behaviour in the common space or to refrain from living together with those who do not practise a way of life similar to his own. In other words, he tends to group in distinct areas in which all the other inhabitants practise the ultra- Orthodox way of life. It should be mentioned that the intolerance is directed against other Jews only since gentiles are not obligated to observe Jewish laws and are not jointly responsible for carrying them out.

However, not all the ultra-Orthodox population lives in segregated areas. The majority is dispersed among the non-religious population in Israel. This raises the question of why the segregation is not complete. Why is it that not all the ultra-Orthodox population concentrates in distinct areas?

The answer to this question is to be found in a concept formed by Martin Buber: 'The line of demarcation' (Simon 1973). Any philosophical idea poses before Man a categorical imperative - a demand to draw practical conclusions as a result of his belief in the idea, in this case, religious belief. Against this categorical imperative, daily life presents various limitations and compulsions on the way to fulfilment of the imperative. The religious individual or group leads himself to a situation or permanent tension between the categorical religious imperative and the constraints imposed by demands, wishes and desires of a human being in a modern world. One who takes upon himself the yoke of the Law undertakes also the task of living by infinitely trying to resolve this tension. There are two uncompromising ways to avoid such a situation. One is to become non- religious; the other is to devote oneself to religious life only. By adopting the latter solution, one minimises the constraints imposed by the world at large and encloses himself in the narrow area of religious law. However, when this Law is practised in a state of almost complete detachment from the outside world, it acquires a special significance of its own. That is the world of ultra-Orthodox Judaism. The decison to live within the discrete ultra- Orthodox boundary, or on the outside among other people, is also the result of drawing the line of demarcation between the categorical imperative and the limited ability to carry it out.

The ultra-Orthodox population, through residential segregation, dissociates itself from the outside world, not only as a result of a wish to create a territorial space in which its own way of life will be dominant, but mainly out of opposition to the non-religious way of life and the wish to dissociate from it (Friedman 1977: 129-45; Poll 1969; Rubin 1972; Mayer 1979). Since the definition of an ultra-Orthodox Jew is based on a way of life, there are no objective statistical data on population that can be measured and that can serve as indicators for the identification of the ultra-Orthodox population. However, the unique way of the ultra-Orthodox population has several spatial expressions which can be mapped and which can indicate the extent of population distribution. Four such expressions are examined here.

Voting   The cohesive nature of the ultra-Orthodox population, its tight social control and religious fundamentalism result in uniform patterns of behaviour, including voting in national elections for a single party, Agudath Israel. This party represents the interests of the ultra-Orthodox population in the Knesset (Israeli Parliament) and in those municipalities in which the ultra-Orthodox population possesses sufficient votes to bring it representation. Mapping those areas which have a high percentage of voters for Agudath Israel will represent the residential areas of this population. The drawback of this method is that the voting areas do not overlap with the ultra-Orthodox residential areas exclusively and in the border areas of the ultra-Orthodox neighbourhoods there are also concentrations of non-Orthodox population. It is difficult, therefore, to determine accurately changes in the lay-out of the ultra- Orthodox areas and notably so in their expansion fronts.

The ultra-Orthodox educational system   The factors set out above also differentiate the ultra-Orthodox education system from the national educational system of Israel, including the State-religious educational system. Ultra-Orthodox children study in autonomous educational institutions which belong to the

103

ultra-Orthodox population. Some of these are state-financed as a result of political agreements, but the state is excluded from any influence on the educational and teaching programmes in these institiutions. Mapping the homes of the children in the ultra-Orthodox educational institiutions marks the areas of expansion of that population. Yearly updating of the addresses of the pupils allows the dynamics of spatial changes in the population under study.

The 'Eruv'   According to the Halacha, a Jew may not carry objects on the Sabbath from his private domain to the public domain and vice-versa, or from one private space to another. In order to permit observant Jews to carry objects on the Sabbath, it is customary to surround the territory in which observant Jews wish to move with a wire (an Eruv). From a religious legal point of view, this creates a space considered to be one continuous area in which transport of objects is allowed. The whole area of Jerusalem is surrounded by an Eruv set up by the Religious Council, a municipal body not recognised by the ultra- Orthodox population as it is part of the secular system of government. As a result, the Religious Court of the ultra-Orthodox community installs an Eruv of its own which spans the territory in which the ultra-Orthodox population resides. Penetration by an ultra-Orthodox population into new territories will cause an extension of the ultra-Orthodox Eruv, and mapping these extensions will show the expansion process of the ultra-Orthodox area (Figure 10.1).

Blocking of streets on the Sabbath   Driving motor vehicles on the Sabbath is perceived as a gross infraction of the Sabbath laws and as a provocation against those who uphold those laws. Therefore, in any place in which the ultra-Orthodox population has grown into a sizable group, it interferes with the movement of motor vehicles on Sabbaths and Jewish holidays. When the ultra-Orthodox population becomes a majority in a given area, it solicits the aid of its political representatives in the City Council in order to block the area to the movement of motor vehicles on an official basis. Figure 10.2 illustrates the blocked zones in Jerusalem on the Sabbath and represents those areas in which there is a complete dominance by the ultra-Orthodox population.

The first two characteristics serve as indicators of the growth of the ultra-Orthodox population in certain areas, whereas the latter two are the result of this growth. Figure 10. 3 illustrates those areas in Jerusalem under Jewish ultra- Orthodox cultural dominance.

Figure 10.1: 'Eruv' Extension in Jerusalem

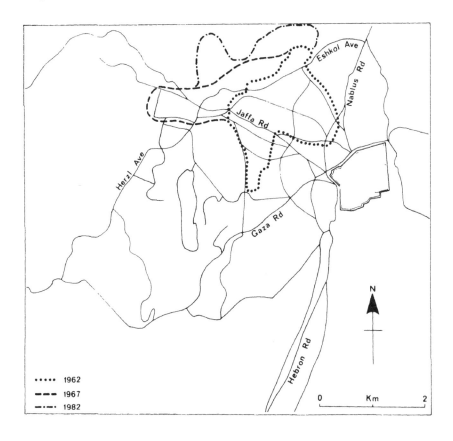

Figure 10.2:   Areas Blocked To Traffic on Sabbath,
                Jerusalem, 1981

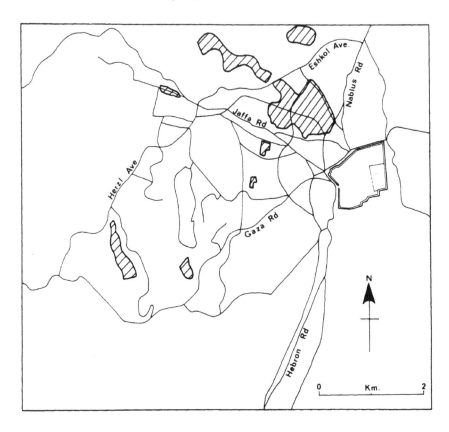

Figure 10.3: Areas With Strong Ultra-Orthodox
Dominance in Jerusalem

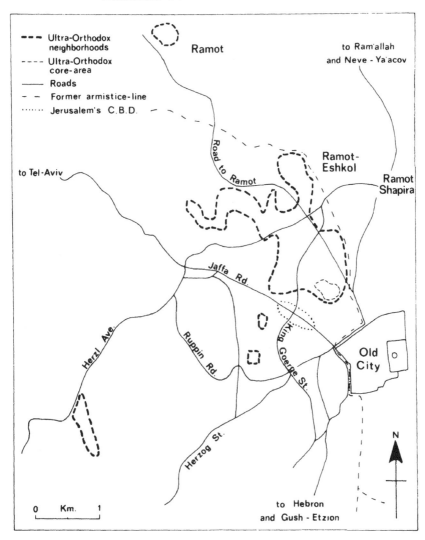

Conflict between ultra-Orthodox and secular

As Israeli society has become more secular and as the Western
way of life becomes more entrenched in Israeli society, particularly
in the consumption of leisure time in the public domain, the
conflict between the ultra-Orthodox population and non- religious
Jews has increased and the tendency towards ultra-Orthodox
segregation has become stronger.

The territorial core of the ultra-Orthodox population is
in northeastern Jerusalem in the area of neighbourhoods established
from the second half of the nineteenth to the early twentieth
century. Out of this core, this population extended its hold
in an ecological process resembling the invasion and succession
processes so well known from American cities (Burgess 1928;
Johnston 1971). This extension was directed towards nearby
neighbourhoods which, until the 1950s were mixed or secular.

The Jewish ultra-Orthodox population has special demographic
and socio-economic characteristics. Its birth-rates are higher,
and its average marriage age and labour force participation lower
than the Israeli average. As a result, its average income level
is also lower and as a community, it possesses a high level of
communal cohesiveness. The significance of this is that the
demand for housing in areas where this population concentrates
is high, but the ability of individual members to obtain housing
on the free market is limited because the economic ability to
compete for the limited available space falls short of demand.

The segregation of the ultra-Orthodox population is not a
spontaneous process but is aided by affiliated religious
organisations and political parties. In this the modern-day
ultra-Orthodox population bears similarities with the general
Jewish population in Palestine at the time of the British Mandate.
Then, the British authorities failed to supply the services
commonly considered normal in a modern welfare state. As a
consequence, the Jewish population fended for itself through the
establishment of voluntary groups to provide a range of social
services which included medical insurance, education, cultural,
employment and union organisation. In the absence of the
appropriate institutions, these roles were performed by the
political parties. The religious parties formed part of this
Jewish organisation and divide into two major subgroups, one
under the aegis of the National Religious Party, a Zionist party,
and the other under the umbrella of Agudath Israel, which is
officially non-Zionist. Another anti-Zionist group believing
in the establishment of a Zionist theocracy and Messianic
redemption of the Jewish state, represents the rest of the
ultra-Orthodox population. Although it does not normally cooperate with the general political system, this group is capable
of exerting pressure and threats on leaders and members of the
religious population.

In the current political structure in Israel, the government
coalition is dependent upon the support of the religious parties.

This is bought by answering the demands of the ultra-Orthodox population for land and public housing and the supply of educational and religious services to the newly-constructed residential districts.

Ultra-Orthodox Expansion

The mechanism for expansion of ultra-Orthodox residential districts has been assisted by communal procedures that express themselves at two levels.

First, combined activity in setting the stage for invasion into new residential areas may express itself in the gathering of information concerning the availability of apartments in the required areas, mutual economic aid in order to provide the capability to purchase apartments, organising to prevent the penetration of non-religious families into areas desired by the ultra-Orthodox population, including the organisation of territorial conflicts to eject undesired land uses from the area. Indeed, the extension of ultra-Orthodox residential areas has been accompanied by conflicts along the advancement fronts of the ultra-Orthodox areas. The aim of these conflicts has been to exclude non-religious populations from these areas by removing secular institutions such as schools or youth clubs and to modify the behaviour in the area by such means as blocking streets to motor transportation on Sabbaths, exclusion of certain forms of entertainment and amusement or the harassment of women whose dress is considered insufficiently modest.

Second, the ultra-Orthodox population has been active in securing housing by the creation of special construction and housing companies. These companies acquire cheap land outside the developed urban areas, building extensively and then offering very cheap housing for their population.

The combined action of these two factors has led to the current spatial organisation of the residential areas of the ultra-Orthodox population. Through the invasion and succession process which occurred in the 1950s and 1960s the ultra- Orthodox population filled the space in the north of the city which is bounded on the north and the east by the armistice line which divided Jerusalem between 1948 and 1967. In the south it was bounded by the CBD and in the west by the edge of the city's built-up area.

As a result of the invasion by the ultra-Orthodox population into non-Orthodox districts and the institutional apportioning of construction for the ultra-Orthodox population, distinct residential districts have been created, the uniqueness of which lies not only in the sociological and cultural traits of the population but also in their physical nature and spatial order.

The ultra-Orthodox conception of Judaism makes a distinction between values and ways of life which must be avoided and those which do not endanger its way of life and can thus be adopted. Thus, the ultra-Orthodox population is not totally estranged from the surrounding world. It is selective in its adoption of technological innovations choosing those which do not serve as vehicles for the transfer of cultural values and characteristics which may endanger its beliefs. Electronic communications devices are commonplace while TV is taboo; the best medical services available will be sought but autopsies are fought.

This duality spills over into the urban landscape. Commercial and business contacts are maintained with non-observant Jews and non-Jews but social contacts are negligible, restricted mainly to ritual gatherings, lest secular influences affect the culture. In social contacts, education and culture, the ultra-Orthodox are restricted to their own segregated areas which include the schools. Study time and leisure time are spent in these districts. All accepted modern forms of leisure are forbidden and this has an important effect on the ease with which the young men (exempt from Army service) can be mobilised as a force for street demonstrations (Freedman, personal communication). Such demonstrations are often organised to help impose the ultra-Orthodox way of life in ever-expanding sections of Jerusalem.

The space available for the activities of the ultra- Orthodox Jew may encompass the whole city during the week and for commercial intercourse whereas on the Sabbath and in social and cultural intercourse, it is very restricted.

Certain unique features also exist in the physical appearance of residential buildings. Israeli building laws require an elevator in a building in which more than four floors must be climbed. As the ultra-Orthodox do not accept the solution of constantly operating automatic elevators on the Sabbath (in contrast to many orthodox residential buildings), the buildings inhabited by the ultra-Orthodox are generally low-rise buildings. Apartments in such buildings will also possess an uncovered balcony for the construction of a succah, a ritual booth used during the Feast of Tabernacles. There are also no TV antennae, cinemas or advertisements in ultra-Orthodox neighbourhoods; buses carrying 'immodest' advertisements and which pass through ultra-Orthodox neighbourhoods may be in danger. On the other hand, the landscape contains many synagogues, yeshivot (rabbinical institutions of study) and batei midrash (religious academies); the large number is partially due to the extreme dichotomies in the ultra-Orthodox community, each subgroup attempting to preserve the traditions of specific East European towns. The commerce of these districts is marked by a plethora of shops selling religious articles and kasher (ritually certified) food under the supervision of the rabbi of the sub-community. Because of the absence of mass advertising media, posterboards and graffiti serve the ultra-Orthodox with community notices of meetings and demonstrations. In addition, the urban landscape contains many clothing shops catering to the peculiar fashions of the community and baby-shops which reflect the high birth rate.

110

THE POLITICAL-GEOGRAPHICAL CONFLICT

In the late 1960s a _modus vivendi_ was reached in the relationships between the secular and ultra-Orthodox populations in Jerusalem, which resulted from the territorial niche which the ultra-Orthodox population had carved for itself in the city. From this niche, the ultra-Orthodox population maintained ties with the secular community on matters and timing of its choosing. On weekdays, the city functions as one system on all matters other than educational, social and cultural ties; on Sabbaths and Jewish holidays, the ultra-Orthodox are confined to their territorial niche and disengage themselves from the secular system.

The spatial situation of the ultra-Orthodox areas changed after the 1967 war. The Israeli decision to extend Jewish suburbs beyond the former armistice line converted the territorial niche into an enclave, bounded on all sides by the secular system. As a result of the construction of suburbs and neighbourhoods north of the city, new roads crossed the ultra-Orthodox areas and created tangential lines and contact areas between the ultra-Orthodox and non-religious populations. The new situation recreated the confrontation between the two concepts that are so far apart.

Moreover, the planners decided to establish a new suburban neighbourhood, Ramot, and a new football stadium north of and close to the ultra-Orthodox neighbourhoods, in the space regarded by the ultra-Orthodox population as their land reserve. The area north of Jerusalem represents the only spatial solution for the extension of the ultra-Orthodox quarters. This blocking of expansion-potentials reawakened territorial conflicts. These conflicts have aleady prevented the construction of the stadium and are now focussed on preventing the use of motor transportation on the road to Ramot on Sabbaths and on the penetration of ultra-Orthodox into Ramot itself. This penetration is accomplished by the organised acquisition of new housing projects for ultra-Orthodox groups.

The attempts to block the Ramot road on Sabbaths, as well as the partly successful attempts to penetrate Ramot have met with the active resistance of the non-religious population of Ramot and have created a tense situation between these groups.

The spatial policy of the authorities in the north of Jerusalem has encouraged the emergence of spatial conflicts between populations. However, the materialisation of these conflicts depends not only on the geographical relationships between the groups but also on the internal socio-political structure of the disunited ultra-Orthodox population which is so sharply subdivided into many groups each competing with the others for hegemony within that population.

These conflicts are, today, the problematic focus of intra-Jewish relations in Jerusalem and they pose one of the challenges for future urban planning.

This paper is based on a study supported by a grant from the Ford Foundation received through the Israel Foundations Trustees and was carried out in the Jerusalem Institute for Israel Studies.

REFERENCES

DOWNS, A. [n.d.]
'The future of American ghettoes', in W. Gorham et al., Urban Processes, (The Urban Institute, Washington, D.C.).

ERICKSON, E.G. [1954]
Urban Behavior, (New York).

FRIEDMAN, M. [1977]
Society and Religion - the non-Zionist Orthodox in Eretz Israel 1918-1936, (Yad Ben-Zvi, Jerusalem), (in Hebrew with English summary).

HERBERT, D.T. and R.J. JOHNSTON (Eds.) [1978]
Social Areas in Cities, (Wiley, New York).

HERSHKOVITZ, S. [1976]
'The Spatial Structure of the Religious Population in Jerusalem, 1959-1969', Studies in the Geography of Israel, 9, 156-74, (in Hebrew).

MAYER, E. [1979]
From Suburb to Shtetl, (Temple University Press, Philadelphia).

MEINIG, D.W. [1965]
'The Mormon culture region: strategies and patterns in the geography of the American West', Annals, Association of American Geographers, 55, 191-220.

MORRILL, R.L. [1965]
'The Negro ghetto: problems and alternatives', Geographical Review, 55, 339-361.

POLL, S. [1969]
The Hasidic Community of Williamsburg, (Schocken, New York).

RAYMOND, E.S. [1970]
'Residential desegregation: can nothing be accomplished?', Urban Affairs, 265-277.

ROSE, H.M. [1971]
'The development of an urban subsystem: the case of the Negro ghetto', Ekistics, 31, 183, 137-42.

ROSSI, P.H. [1955]
Why Families Move - A Study in the Social Psychology of Urban Resdiential Mobility, (Free Press, Glencoe).

RUBIN, I. [1972]
Satmar: An Island in the City, (Quadrangle, Chicago).

SIMON, A.E. [1973]
   'The Line of Demarcation' - Nationalism, Zionism and the
   Jewish-Arab Conflict in Martin Buber's Thought, (Arab and
   Afro-Asian Monograph Series, #12, Giv'at Haviva), (in
   Hebrew).
TIMMS, D.W.G. [1971]
   The Urban Mosaic: Towards a Theory of Residential
   Differentiation, (Cambridge University Press,
   Cambridge).
WILSON, B.R. [1967]
   'An analysis of sect development', in B.R. Wilson (ed.),
   Patterns of Sectarianism, (Heinemann, London), 22-45.
WILSON, B.R. [1972]
   'Religion in secular society', in R. Robertson (ed.),
   Sociology of Religion, (Penguin, Baltimore), 152-62.

PART III: TERRITORIAL PERSPECTIVES ON PLURALISM

Chapter 11
THE DILEMMA OF NATIONS IN A RIGID STATE STRUCTURED WORLD
David B. Knight

INTRODUCTION

First, imagine the total removal of all existing international political boundaries and then, thinking of all distinctive large population groups as regionalisms, impose new political boundaries around those populations. Two things would stand out: the location of the new boundaries would be quite different from currently existing ones and there would be hundreds of new political units. Perhaps it is only wishful thinking to consider such a total reorganisation of political space, although to some writers, who see bigness as the cause of all forms of human misery, hundreds of new 'states' might represent the salvation of the world (see Kohr 1957; Schumacher 1974). But would a world of only small (many of them mini) states necessarily be good? Diversity is one of the key stimulants to successful societies, not uniformity. The dilemma to be faced in any such redrafting of international boundaries would be, what scale of population is 'proper' to enjoy a territory and government of its own? There is no easy answer to this question.

People have attachments to several levels of territory, but the dominant structure in today's world is the state and so, at one level of generalisation, all people on earth are defined by state levels of large group identities since the areal extent of the approximately 160 states (politically sovereign countries) are said to define the people who live within their bounds. Unfortunately, however, this statement has more basis in theory than fact, for many groups do not identify with the states within which they live. Some of these people belong to groups split by international political boundaries which do not conform to cultural distributions. Many others are minorities within their states' populations and may be seeking either a degree of political autonomy within the existing states or even total separation to form new states of their own.

Of the many groups of people which exist, most have a sense that they are unique in many respects, not least in having a sense of separateness from all other groups. For many, the

114

primary group identity today is referred to as the 'nation' with
the political expression of the nation being its 'nationalism'.
Our world is made up of a mosaic of nations which form one (but
key) element in the socio-cultural, political and economic
international web of interdependent regionalisms. In recent
decades we have seen the near worldwide application of the concept
of self-determination, as numerous former colonial territories
have achieved political independence. The resulting many
redefinitions of territory should not be seen as a signal that
we have arrived at the end of any kind of 'evolutionary ladder'
of territory formation, however, for we should fully expect
territorial changes to continue. We can thus ask the question:
by considering existing states from the perspective of group
territorial identities, can we get any clues as to what some
future changes might be, that is, beyond those which might occur
if there is a major war between world powers or, even, other
members of the world community?

        This question is addressed by focussing on some general
theoretical thoughts before relating them to one country, Canada,
in which a variety of group/territorial processes are operating
and which threaten to bring about change. Some general comments
about the international significance of group/territorial
identities are then made. The perspective adopted here is
territorial, with certain political, social and economic elements
knowingly remaining only implicit.

TERRITORIAL ATTACHMENTS AND TERRITORY[1]

        We live out our lives within a hierarchy of territorial
organisations. At the core is the home but much of our lives
relate to territorial organisations which are far removed from
the personal level and yet which often influence us directly.
In most western countries, for instance, rural townships and
counties, as well as towns and cities are delimited, and within
their bounds certain regulations exist which may differ from
those found in neighbouring municipalities. At a higher level
of territorial organisation, at least in federations (as in West
Germany, the United States, Canada and Australia), states or
provinces exist and, again, different sets of regulations may
exist within them. And, of course, the various levels of
governmental organisations are grouped together under national
governments, with sets of rules which pertain to the total state.
There is a hierarchy. Indeed, all human territorial organisations
– be they political, social or economic – generally involve the
concepts of rank and hierarchy, which vary in complexity according
to changing scales and perceptual frameworks. Further, we are
also influenced by certain cultural values. For the moment,
however, it is the perceptual perspective on territorial ties
that is to be stressed. Clearly, we do have ties to different
scales of territory and we can operate at several levels of

abstraction at any one time - from personal to small group to parochial localism to a broader regionalism to a nationalism and even to an internationalism (see Lowenthal 1961; Saarinen 1976). We have the astonishing ability to shift perspectives in our minds and thus change levels of abstraction. Our 'sense of place', as defined at any one of these levels of abstraction, has a territorial component. The focus here is on the national level and lower level regionalism.

Jean Gottmann (1973: 15) suggested that territory '. . . is replete with inner conflicts and apparent contradictions'. We can declare that the latter results from man's beliefs and actions, for territory itself is passive. Knight (1977) showed how written and spoken expressions of power and meanings of Canadian territory found expression within the political system as politicians and newspaper writers manipulated their notions of place and space while they tried to resolve a highly contentious locational issue. Territory is not, it becomes, in the sense that over time people come to 'see' meaning in or 'obtain' meaning from territory and its landscape and may fully believe in the latter as a living entity (Eliade 1961; Tuan 1974; 1979). While such meanings are all psychologically and culturally based (and a particular territory can mean one thing to one group and quite another to a different group), any group's cultural ecology and spatial patterning can and does find expression in the way space is structured and how land is used, and landscapes as perceived by the occupants can have powerful symbolic links to a group's territorial identity (see Tuan 1974; English & Mayfield 1972: 3-319; Jordan & Rowntree 1982; Lowenthal & Bowden 1976; Mikesell 1968; Knight 1971; Kliot 1982; Chapter 14, this volume).

In addition to symbolic links, at a group level, we can ask: what does territory provide? Gottmann (1973: 17-24) has suggested that territory, when delimited with a system of government (the type does not matter) that has effective control over it, provides both security and opportunity for those who live within its bounds. From the one perspective, security, there is an inward-looking Platonic ideal, with isolated territory, self-sufficiency and a stress on security. From the other, opportunity, there is the Aristotelian position that there should be an outward perspective, with participation in a larger (perhaps international) system. Security as an isolated community; opportunity as part of a larger whole. There is stress caused by these two contradictory dimensions of territory, with elements of both always being present, perhaps with states sometimes being further along the continuum towards one extreme or the other at different times.

The state, a legal and physical entity, is a bounded container for the contents of a particular portion of earth space, and includes the land, people, resources, and means for communication and movement. Also through an effective system of government, according to Greer and Orleans (1964: 810), the state is the chief custodian of overall social order; it is the monitor,

116

comptroller, arbitrator. There are six basic characteristics to any state, according to Glassner and De Blij (1980: 43- 44): land territory, permanent resident population, government, organised economy, circulation system and, most important perhaps, sovereignty and international recognition. The modern state, as the dominant unit which divides territory, became, to use John Herz's phrase, a 'hard shell' towards its external environment, although Herz noted too that technological developments have enabled belligerents to leap over or bypass the traditional hard-shell defence of states (Herz 1976).

LARGE GROUP ALLEGIANCE

'National' communities generally desire to have the bounds of their political space coincide with the areal distribution of the distinctive population so that their government can best protect the community's interests. This is a basic underlying thrust of all nationalisms. As the first of then possible elements in nationalism, the nationalist historian Boyd Shafer (1966: 3-4) listed:

> 'a certain defined, even if vaguely, territory
> or land, whether this be large or small,
> possessed or coveted. This land is said to
> belong to the nation, and nationalists, in
> varying degrees, think of it as their own
> and are devoted to it.'

Elsewhere, Shafer suggested that the nation becomes the dominant social grouping because men determined that they might live in groupings larger than family or tribe, and that they could not comprehend an international or universal state (Shafer 1955: 97; see Kohn 1944; Shafer 1972; A.D. Smith 1971; G.E. Smith 1979; Knight 1982).

Whereas people once defined their territory, in time territory came to define the people who lived within it. Thus, for instance, as the historian Maine put it for Englishmen, 'England was once the country in which Englishmen lived: Englishmen are now the people who inhabit England' (cited in E. Jones 1966: 56). As people in different parts of a territory developed a sense of national unity there was, as Shafer has noted, 'the disappearance of other unities and distinctions, such as those of privilege and province (Shafer 1955: 105). It was in the eighteenth century that people in the middle classes in western Europe and North America came increasingly to identify with the feeling that the nation belong to them. Gottmann (1973: 72) has described how, during that century, a great doctrine emerged which called for freedom of all men and the idea 'was advanced that every nation should be allowed to elaborate its own system of laws (see Gottmann 1975: 29-47). The American and French revolutions led to the acceptance of the idea of national sovereignty over the inhabited territory, where jurisdiction was claimed in the name of the body politic. The course of world history was subsequently changed, and we find that during the past 200 years there have been hundreds of territorial changes. Many of these changes resulted from military might, or treaty arrangements following conflict, but dozens of other changes have occurred because of the claiming of colonial territories and the more recent transition from colonial status to political independence. The latter transition of territorial definition resulted from the idea of self-determination being transformed from idea to process.

Territorial changes are still occurring. We can place these changes into two broad categories: unification or separation of territories. In the present era, amidst an ongoing search for a new international order, there is, among many other dynamic processes, the move towards new supra-national unifying forms of linkages such as the European Community or the New Zealand-Australia free trade area (with, at least in the minds of some Australians, a political union somewhere down the line!) (Feld and Boyd 1980). In contrast, but also part of the 'winds of change', a different force has developed (perhaps as a neutral extension of the self-determination thrust by former colonial territories), namely, the rise of regional groups within multi-national states which desire to recover past identities through the achievement of independence which would involve the separation of territories. In a sense, with the latter tendency, we are seeing the rise once more of 'provincialisms' which Shafer suggested normally die following the development of nationalisms.

A 'provincialism' or regionalism becomes evident only at certain levels of abstraction, at certain scales of generalisation. Britain, for example, is often presented as a good illustration of a 'nation-state' but, of course, four major culture groups exist within the state, with each group possessing attributes of nationhood, with a measure of distinctive identity which is associated with a particular territory. At one scale of generalisation, the distinctive consciousness that exists within Scotland, for instance, is only a regionalism or provincialism, if viewed from a Britain-wide perspective, but, naturally enough, Scots nationalists claim it to be much more. For the latter their 'regionalism' is a 'nationalism' which is rooted in place, that place being the territorial extent of Scotland, wherein the Scots nation and state coincide (Alexander 1981; Agnew 1981).

Scotland within Britain is not unique, for similar culturally based, or ethnic regionalisms also exist elsewhere (Antipode 1980; Williams 1976; Morgan 1981). However, by no means can we call all such ethnic regionalisms 'nationalisms', even though cultural minority groups may nevertheless have powerful senses of unity. At a minimum we can talk of group consciousness which is generally tied to specific places or regions. Of particular note, such expressions of group consciousness can exist (at one and the same time) alongside feelings of attachment to broader regional groupings, involving, for instance, the symbolic linkage of all groups which exist within the same national territory and, at a different level, to some kinds of interstate organisations (such as the European Community). This ability to have an identification with the state society and also interstate, perhaps world regional, organisations parallels the process internal to the state, where people can have the identification with their regional, sub-group and local attachments as well as with the 'nation'. Even so, we still live in a world, as the Wilsons (1965: 41) put it nearly four decades ago, where:

> 'A civilised man who puts his town or his family before his state would be judged wrong by his fellows, though one who put his state before some wider group is still usually judged to be right.'

Clearly, the state remains as the principal focus for most large group allegiances. However, within any state, when a regional (especially ethnic group consciousness finds explicit expression then there is the potential for conflict.

Connor (1973) has referred to ethnic regionalisms which desire a degree of political autonomy (or even independence) as ethnonationalisms. Such identities are rooted in a variety of emotional, historical and political-economic realities, as perceived and defined by members of the national group itself. Where ethnonationalisms exist they seem to find greatest expression in peripheral portions of states. Their existence undoubtedly has some relationship with unevenly distributed processes of economic development, even to the point where the regional hinterland locations can be thought of as 'internal colonies' of the core areas and their states (Hechter 1975). As the elites of regional ethnic groups formulate their separatist political ideologies, they use such phrases as ethnic or cultural integrity, uniqueness, minority status, the desire for national freedom, and the right to independence. Anthony Smith (1979: 22) feels that an ethnic separatism, in seeking a separate political existence, hopes to achieve the restoration of their 'degraded' community to its rightful status and dignity.[2] Of course, it follows that for an ethnonationalism to achieve a separate political existence there may also be territorial separation. The threat of ethnic separation exists to varying degrees in many modern Western countries (including Canada, Britain, France, Belgium, Spain, Switzerland, the United States and Italy (Williams 1980; 1982). We can observe the phenomenon elsewhere in the world, too, with some efforts to achieve a separate existence having resulted in terrible losses of human life, as in Biafra, Eritrea and Bangladesh (see Hall 1979; Dofny & Akiwowo 1980; Whebell 1973).

SELF-DETERMINATION AND LEGITIMACY

Why do ethnoregionalisms, or ethnonationalisms, seek independence? Rene Levesque, Premier of Quebec Province in Canada, has declared (1978: 110) that 'self-determination is an absolute necessity.for the growth to maturity of a society which has its own identity.' By saying a 'society' having its own 'identity' Levesque raises the question of whether he is referring to a nationalism or to something less than that. If it is to a nationalism then one of the many attributes of nationalism is having or coveting a territory (Burghardt 1973). If nationalism can include the latter, sovereignty, in contrast, means actually having and controlling the territory - and this is just what is desired by any nationalist. Perhaps we should ask what is meant by self-determination? Rupert Emerson put it simply: 'self-determination constitutes formal recognition of the principle that the nation and state should coincide.' He added: ' the plain fact is that the stage structure derived from the past only occasionally and accidentally coincided with the national make-up of the world.' (Emerson 1971: 294). Plural or multicultural societies created in part by colonial boundary delimitations thus are, for all practical purposes, the order of the day.

While self-determination may be a sought after goal for the regional group, notions of national integrity will forbid nationalists of the existing state from acknowledging the self-determination thrust by the minority as being legitimate. This occurred in Canada during 1979-80 when the federal Prime Minister and most provincial premiers stated that they would not negotiate sovereignty assocation with Quebec if the 'yes' vote won in the Quebec referendum that was held on the issue of separation of Quebec from Canada. Such a reaction was not at all surprising, for all people generally recognise self- determination 'elsewhere', that is, outside their own territory; to do otherwise is to acknowledge the 'rights' of separatists. For example, the French may support Quebecois self-determination but not the claims of Bretons; the Russians support self-determination (or national liberation?) for groups in many parts of the world but not the claims of the numerous minority groups within the 'national territory' of the U.S.S.R.

In reality, an areal match between nation and state rarely exists, thus there are many plural societies. Always implicit in a plural society is an element of strain, a 'we' versus 'they', and an important question always is 'who rules whom?' In some plural societies, the regional differences are given political recognition through a federal structure, of which several varieties exist. Some exist in name only because of top-down dominance, whereas others give major emphasis to the regional components of the system at the expense of a strong central authority.

CANADA AND ITS REGIONALISMS

Canadian provincial and federal authorities have wrestled with the issue of balance between regional and central power ever since 1867 when the Canadian federation came into being. A federal system of government can permit considerable regional autonomy (as in the case of Canada), but most parts of the federation, perhaps any federation, will always want still more power. However, there is a limit to which regional accommodation can go before disintegration of the state results. From this we can ask: to what extent can regional identities and regional economic power be encouraged to flourish at the expense of the national state's very existence. Equally, how far can the central government increase its own powers at the expense of the regional authorities? Clearly, for there to be a true federation, there needs to be a delicate balance maintained between central and regional authorities, with everyone remembering A.V. Dicey's warning (1939: 602) that federalism rests on the psychology of the people of the political units involved desiring union without desiring unity.

Canada, for all of its successes as a federal state, has been a country with socio-cultural, economic, political, and also territorial cleavages. From the outset of British North America's separate existence following the Revolutionary War in the territories to the south, the several societies located across the land each evolved in their distinctive ways, with each attaining quite different perspectives on 'Canada', these being based part on relative location within the expanding national territory. Territorial cleavages occurred, as Prince Edward Island and New Brunswick were carved out of Nova Scotia, Upper Canada (later Ontario) out of western Quebec, and other provinces out of prairie lands. Regionalism was well rooted before confederation in 1867. Ontario, Quebec, New Brunswick and Nova Scotia were linked by the federal system in 1867, and were joined by Manitoba in 1870, British Columbia in 1871, Prince Edward Island in 1873, Alberta and Saskatchewan in 1905 and Newfoundland in 1949; Yukon and Northwest Territories stand alongside as federal territories. The country thus stretched from the Atlantic to the Pacific and from the United States border to the Arctic archipelago, ranking second to the U.S.S.R. in territorial extent. Notwithstanding its size, Canadian society can still be described, to a large degree, as being comprised of regional societies which are often quite parochial in sentiment and outlook and particularist in spirit.

The Canadian challenge has always been to seek and weld an identity that transcends these several parochial societies which are found within different parts of the national territory. The Federal Government thus has as one of its principal purposes the nurturing of a national identity and will. This apart, the boundaries of the regional societies are only vaguely known (Merrill 1968: 556-68; Clark 1962: 23-47; Ray 1971; Harris 1979;

121

Task Force on Canadian Unity 1979). Many Canadians obviously identify with their province, but there may also for some be a broader identity which transcends the more parochial provincial identities (Morse 1980). Sometimes the leaders of the four Western provinces and the four Atlantic provinces act or at least speak in concert as they seek a redress of actual and especially perceived injustices by the Federal Government, with most contemporary federal-provincial conflict hinging on the issue of who controls 'provincial' resources.

Increasingly, in Canada, the 'periphery' is flexing its muscles against what it perceives to be an insensitive 'centre' (McCann 1982; see Gottmann 1980). There have even been moves by minority groups to seek the West's separation, or Newfoundland's separation, or, especially Quebec's separation from Canada (Williams 1981; Pratt & Stevenson 1981; Knight 1982). An important point to note is that whereas Quebec separatism is a movement with deep cultural-historical and territorial foundations, the other regional separatist movements are essentially economically based, even though one cannot deny the existence of elements of western or eastern Canadian cultures and regional identities that are considerably more than economic in nature. Given a strong economic thrust, such separatist movements are dangerous for the continued life of the Canadian state. It seems certain that if Quebec 'nationalists' had been able to exploit the feelings of alienation which had an economic base, then the Quebec electorate would have undoubtedly voted for separation. Since 60 per cent of Quebecers went on record as having the will to remain in Canada, Premier Levesque has acknowledged democratically that he did not have the power to lead Quebec to independence by whatever name. At the same time, he and his Parti Quebecois government have made it quite clear that they still want Quebec's separation from Canada and they have announced that the next provincial election will be run on that issue. For, if federalism rests on the psychology of the people, then Richard Hartshorne's conclusion that the most powerful centripetal force for any state is simply the people's belief that the state should exist (Hartshorne 1950).

Interestingly, in the pre-referendum debate within Quebec, it was implied that Quebec, as the 'state', represented the embodiment of the Quebecois nation, that there is a 'national' Quebec culture. To 'belong' to Quebec therefore one also had to belong to the 'nation'. However, many Quebec residents, Francophones or of non-Francophone minorities, also have an allegiance to a greater entity called Canada. They thus have allegiances to both Quebec and Canada, the latter being a level of abstraction to which the Quebecois separatists cannot develop an attachment.

But the Quebec case in Canada does not stand alone. The Parti Acadien has as its ultimate goal an autonomous Acadian province carved from the northern half of New Brunswick where French-speaking Acadians - who see themselves as quite distinct from neighbouring Quebecois - form the majority of the population (Griffiths 1973). There are small numbers of people elsewhere in Canada who feel that their politico-territorial aspirations also deserve to be met. Indigenous peoples across Canada, as in other countries, have experienced an awakening of 'older forms of territorial consciousness' (but which are now framed within European strictures) as they seek to protect their cultural and economic well-being (Raby 1974; Berger 1977; Office of Native Claims 1978). In the eastern portion of the Northwest Territories, for instance, Inuit (Eskimo) form the majority. Should that majority be permitted to secede from Canada if it so desired, perhaps to link up with Greenlanders who are currently in the early stages of breaking away from Denmark? Such a thought is far-fetched for the moment. Nevertheless, the Inuit, through the Inuit Tapirisat of Canada (I.T.C.) since 1976, and most notably through their voting in a Territories-wide referendum in April 1982, have made clear their claim to the right of 'self-determination' (see Brody 1975).

In another example, the Dene of the Northwest Territories see themselves, too, as a distinct people and refer to themselves as a 'nation'. In contrast to the Parti Quebecois, the Dene have called for 'independence and self-determination within the country of Canada.' While the Dene and Inuit and other Native Peoples in Canada see a future in Canada, the 'hard-core' Quebec separatists - or nationalists - do not. This latter issue as to whether to call a person a 'nationalist', 'separatist' or 'coloniser' depends clearly upon one's perception.

We must also remember that there exist other, but not so clearly ethnically-based, expressions of sectionalism in Canada which could also some day lead to territorial separation. For instance, in Nova Scotia, many Cape Bretoners want to have their island become a separate political entity as they seek to achieve a fuller development of their 'selfhood'. And in the West, a political party based on a separatist philosophy won one seat in the Alberta Provincial Legislature in a 1981 by-election. Although the party gained no seats in the Alberta general elections in 1982 it captured more than 10 per cent of the vote. From these additional examples, it can be seen that sectionalist threats can be caused by a wide variety of socio-cultural and economic factors as well as distinct parochial identities.

It is perhaps wise here to establish clearly what is meant by sectionalism. By this is meant the situation when regionalisms find expression to the point where regional political interests are held to be more important than concerns involving the whole state. When sectional interests are carried to the extreme, territorial separation becomes a distinct possibility, so sectionalist tendencies need to be subdued to the point that they do not threaten the very existence of the state. Reformulations of power sharing may be enough to accommodate pressures for change but perhaps territorial restructuring may be needed in some cases.

TERRITORIAL RESTRUCTURING?

Canada's political boundaries took a long time to evolve
(Nicholson 1979). But why should we assume that the end point
has been reached?

First, as geographers know, there are numerous ways to
regionalise any country. Second, most methods have had an
economic framework. Emphasis must also be placed on expressions
of regional consciousness and senses of group within states.
Third, inherited political bounding of territory generally does
not reflect contemporary socio-cultural and even economic
realities. Fourth, established 'communities of interest' will
always want either to retain or expand 'their' territory. It
is not appropriate to develop these points here, although they
underlie some of what is stated below about the possible
restructuring of Canadian territory.

If Quebec separates, with whatever political bounds, Canada
would soon break up. An earlier suggested series of scenarios
for the 'Balkanisation' of Canada stimulated considerable
discussion (Figure 11.1) (Knight 1975). Without elaborating,
let us briefly turn to some other thoughts (Figures 11.2 - 4).

If Quebec secedes, we might suggest that northern Quebec
be retained by the Federal Government as per the boundaries
established in 1867, or revised in 1927. This suggestion might
have had validity prior to the huge northern Quebec James Bay
hydroelectric power developments in the 1970s and 1980s, but now
Quebec clearly can identify its effective control of the region
(and effective control of territory is one of the ways in which
a territory can be claimed). 'Canadian' corridors/strips of land
through Quebec linking the Maritime provinces with Ontario would
be interesting but highly questionable given both the nature of
the terrain and especially the assumption that an independent
Quebec would be willing to give up such land. Given the unlikelihood
of such corridors, the Maritimes clearly would be separated from
the remainder of Canada. Would there be a union of the three
Maritime provinces and Newfoundland? Or might old rivalries
persist so that separate politically independent states would
be created? Or might all or parts of these four provinces simply
become part of the United States?

What of the West? Might old notions of a Prairies' union
be revived? Or might Alberta and British Columbia link up and
become independent, leaving Manitoba and Saskatchewan, and
probably also Ontario, to eventually become parts of the United
States, instead of trying to succeed as land-locked states? Or
would Yukon and British Columbia unite? Might Indian groups,
such as the Nishga, secede from British Columbia to become
mini-states? With increasing economic protectionism across Canada
and with a failed federal system, could each of the provinces
and territories simply become politically independent states,
but with the same boundaries as present? Would they be viable
states? How would the U.S. react to any such changes?

Figure 11.1: Scenarios For Canada's Future

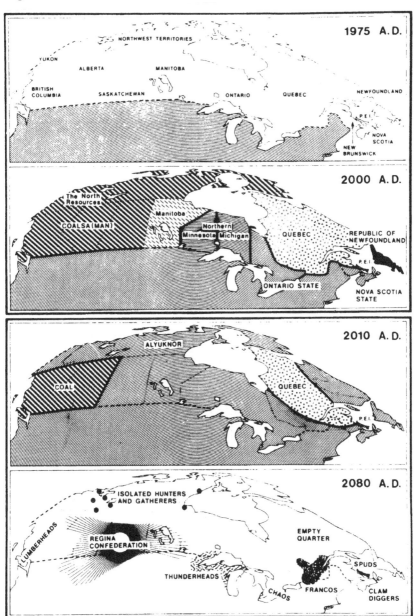

Figure 11.2: Some Possible Secessions?

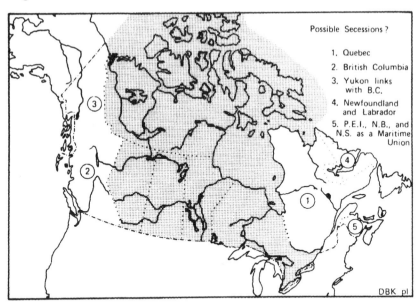

Possible Secessions?

1. Quebec
2. British Columbia
3. Yukon links with B.C.
4. Newfoundland and Labrador
5. P.E.I., N.B., and N.S. as a Maritime Union

DBK pl

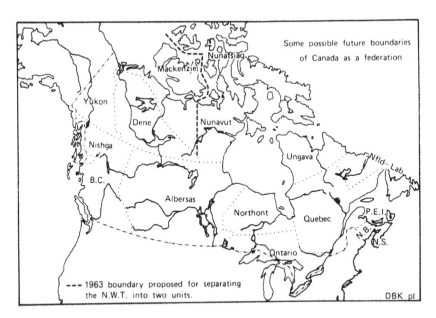

Some possible future boundaries of Canada as a federation

--- 1963 boundary proposed for separating the N.W.T. into two units.

DBK pl

Figure 11.3: Some Possible Future Boundaries of Canada as a Federation

Given that the creation of the three Prairies provinces was
only one of the many proposed ways of politically dividing the
West, should we not think of completely redrafting those boundaries
to fit contemporary realities better? Or should not the British
Columbia/Alberta and Manitoba/Ontario boundaries be redrafted
so that communities linked socio-economically can be united
within the same political units, rather than being split as is
the case now? Should not the distinctive communities of northern
Ontario be given their own separate province? Given that cultural
distributions along the Quebec/Ontario boundary are today quite
different from what they were in 1867, should there not be
boundary adjustments made so that political divisions would
better mirror cultural distributions?   Most of these changes
will not happen, certainly in the short term, simply because
'communities of interest' - the provinces - will always want to
protect their territories.   Indeed, Quebec province actually has
a document which deals with the whole issue of Quebec's territorial
integrity (Dorion 1968; Burghardt 1978; 1980).   Clearly, once
entrenched, territorial bounds are difficult to alter.

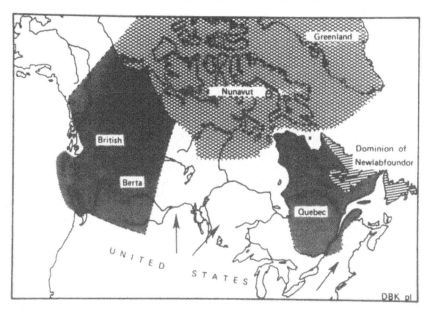

Figure 11.4:   Some Future Possible Territorial Changes

REGIONAL AND NATIONAL IDENTITIES

Canadian provincial and federal authorities recently wrestled
with how to rewrite the British North America Act as a constitution,
and the process still continues as ways are sought to permit the
survival of the federal system with its many regionalisms (Knight

1982; McWhinney 1982). Diversity <u>within</u> unity is possible, notably within a federal system. Federalism is a compact which permits a degree of separateness but 'a federal union depends upon some agreement to differ but not to differ too much' (Wheare 1963). Wreford Watson (1968: 40-41) has declared that 'even in terms of Canada as a whole, in the very context of Canadian unity, diversity exists. . . . the identity of Canada lies in respecting its several identities.' A statement appearing in the so- called 'Beige Paper' of the Quebec Liberal Party during the period just prior to the 1980 referendum noted:

> 'We refuse to choose between Quebec and Canada, as if one should necessarily exclude the other. On the contrary, we choose Quebec AND Canada because each one needs the other to fulfil itself. We are convinced that one can, at the same time, be an authentic Quebecer and a true Canadian" (Quebec Liberal Party 1980: 139-40).

This last sentence can be rewritten for Newfoundlanders, Nova Scotians, Ontarians, British Columbians, the Dene, the Inuit and all the other distinctive groups across Canada. To make this real is the major challenge, whereby regional identities, some of which are also sub-nationalisms, can find full and legitimate expression within the federal system of government.

CONCLUDING THOUGHTS

We should not consider the Canadian search for a new order as unique. It is not. Numerous states are faced with problems of actual or potential disunity because of plural societies. Gottmann (1951, reprinted in Jackson 1964: 27) suggested many years ago that 'the most stubborn facts' in relation to the political partitioning of the world 'are those of the spirit, not those of the physical world,' and that 'one of the most stubborn facts of the spirit remains nationalist feeling - at different scales.' Does this mean that <u>any</u> 'nation' <u>necessarily</u> has the right to independent statehood? If yes, then numerous currently structured states would have to be broken up. But many existing governments (as with those in the Organisation of African Unity), seeking to maintain the status quo, have agreed among themselves that international boundaries, especially those derived from colonial periods, must be accepted as fixed.

In the United Nations, self-determination generally has been accepted for former colonial territories, not peoples within parts of existing states. Not only may any state <u>not</u> invade or otherwise violate the territory of another state but, also, it is <u>not</u> legitimate for people who form a minority within a national territory to seek self-determination (see Nincic 1970: 219- 59;

Emerson 1971; Bucheit 1978; Arangio-Ruiz 1974: 1-13, 131-41; Alexander & Friedlander 1980). In thus accepting the status quo, national governments can deny 'legitimate' claims to independent statehood by groups which now form minorities within their states.

But minorities of all kinds (cultural, ethnic, linguistic, religious and racial) are ever present in our political world (Ashworth 1977; 1978; 1980; Wirsing 1981; Kunstadter 1967; Banks 1976). Some minorities find political expression of their 'selfhood' permitted through a variety of means, such as consociational democracies which permit coalitions or power sharing, minority vetoes, proportional representation, or perhaps some degree of regional autonomy (McRae 1974). In all too many parts of the world, however, full or even partial political recognition and expression of group identities are denied. And so we return to the question about future changes posed at the outset. Will there continue to be changes in the world's international political system as a result of group territorial identities? If current evidence is anything to go by, the answer is yes.

Large group identities - some being 'nations' - will continue to cause conflict, with territorial consequences. Some instances of changes or anticipated changes have already been cited. Many others can be noted. In the Middle East, the world has witnessed the capturing of new territories by Israel as it seeks 'secure borders' but also the daring relinquishing of some of these territories in the quest for recognition, peace and stability (Israelis Reply 1973; Ministry of Foreign Affairs 1982; Harris 1981; Hale 1982). At the same time, there is an anguished cry for Palestinian autonomy and statehood. In Syria, Iraq, Iran and Turkey a similar cry comes from the Kurds (Chaliand 1980). In North Africa, the issue of self-determination has been blatantly ignored in Western Sahara (Franck 1976). In southern Africa, demands continue and pressures mount for majority rule in Namibia and South Africa (Best & De Blij 1977; Brohman & Knight 1981; Christopher 1982). In the South Atlantic, territory recently changed hands forcibly for a short time in the name of 'national sovereignty', with the British justifying its retaking of the Falkland Islands by citing the islanders' right to self-determination. In India, the federal government faces a serious threat by Sikh fundamentalists who seek virtual autonomy in an enlarged Punjab state. This list can continue on and on, although not all instances are accompanied by violence (see Antipode 1980; Boal & Livingstone, this volume; Patrick 1976; Polyviou 1980; Jacobs 1980).

Undoubtedly, as in Canada for the moment, there will be many states which will be able to accommodate societal divisions by constitutional change and perhaps the restructuring of territory (see Adejuyigbe 1977). But, as indicated in this paper, there are potentially numerous groups in many states which could pose secessionist threats. In the short term, it seems likely that most secessionist tendencies will be held in check, in part because of international condemnation of the idea of splitting existing sovereign states - 'if it can happen there it can happen here!' However, it also seems likely that tensions caused by the dilemma of how to give full legitimate expression to 'nations' in our currently rigid state structured world will sooner than later necessarily lead to further fragmentation of political territory, possibly with great loss of life (Brunn 1981). Such is the power of nationalism, where there is the will.

ACKNOWLEDGEMENT: The author gratefully acknowledges the support of the Dean of Graduate Studies and Research, Carleton University, and also the Social Sciences and Humanities Research Council of Canada for a Leave Fellowship and Research Grant.

FOOTNOTES

1. Some points raised in this and the next two sections are expanded in Knight (1982b).

2. An American anthropologist, with respect to problems with terminology, notes that 'ethnic divisions' in their [Third World] societies are 'tribal', those in ours are 'ethnic' (Cohen 1978: 384)

REFERENCES

ADEJUYIGBE, O. [1977]
    'Ethnic pluralism and political stability in Nigeria', in L.J. Evenden and F.F. Cunningham (eds.), Cultural Discord in the Modern World: Geographical Themes, (Tantalus Research Ltd., Vancouver, B.C.).
AGNEW, J.A. [1981]
    'Political regionalism and Scottish Nationalism in Gaelic Scotland', Canadian Review of Studies in Nationalism, 8, 115-29.
ALEXANDER, Y. and R.A. FRIEDLANDER [1980]
    Self-determination: National, Regional and Global Dimensions, (Westview Press, Boulder).
ALEXANDER, A. [1981]
    'Scottish Nationalism: Agenda building, electoral process and political culture', Canadian Review of Studies in Nationalism, 7, 372-85.
ANTIPODE [1980]
    Special Issue on Northern Ireland, 12(1).

ARANGIO-RUIS, G. [1979]
   The United Nations Declaration on Friendly Relations and
   the System of the Sources of International Law, (Sijthoff
   and Nordhoff, Alphen aan den Rijn).
ASHWORTH, G. [1977, 1978, 1980]
   World Minorities, 3 vols., (Quartermaine House Ltd.,
   London).
BANKS, D.J. (ed.) [1976]
   Changing Identities in Modern Southeast Asia, (Mouton, The
   Hague).
BERGER, T. [1977]
   Northern Frontier, Northern Homeland, (J. Larimer and Co.,
   Toronto).
BEST, A.C.G. and H.J. DE BLIJ [1977]
   African Survey, (Wiley, New York), 338-402.
BOAL, F.W. and D. LIVINGSTONE [1983]
   'An international frontier in microcosm: the Shankill-
   Falls Divide, Belfast', in N. Kliot and S. Waterman (eds.),
   Pluralism and Political Geography - People, Territory and
   State, (Croom Helm, London), 138-58.
BRODY, H. [1975]
   The People's Land, (Penguin, Harmondsworth).
BROHMAN, J.A. and D.B. KNIGHT [1981]
   'Some geopolitical aspects of the conflict in Namibia/South
   West Africa', in A.D. Burnett and P.J. Taylor (eds.),
   Political Studies from Spatial Perspectives, (Wiley,
   Chichester), 489-513.
BRUNN, S.D. [1981]
   'Geopolitics in the shrinking world: A political geography
   of the twenty-first century', in A.D. Burnett and P.J.
   Taylor (eds.), Political Studies, 131-56.
BUCHHEIT, L.C. [1978]
   Secession: The Legitimacy of Self-determination, (Yale
   University Press, New Haven).
BURGHARDT, A.F. [1973]
   'The bases of territorial claims', Geographical Review,
   63, 225-45.
BURGHARDT, A.F. [1978]
   'Canada and secession: some consequences of separation',
   in R.M. Irving (ed.), Readings in Canadian Geography,
   (Holt, Rinehart and Winston, Toronto), 10-16.
BURGHARDT, A.F. [1980]
   'Nation, state and territorial unity: a trans-Outaouais
   view', Cahiers de geographie du Quebec, 24, #61, 123-
   34.
CHALIAND, G. [1980]
   People Without A Country: The Kurds and Kurdistan, (Zed
   Press, London).

CHRISTOPHER, A.J. [1982]
'Partition and population in South Africa', Geographical
Review, 72, 127-38.
CLARK, A.H. [1962]
'Geographical diversity and the personality of Canada',
in H.L. McGaskill (ed.), Land and Livelihood, (New Zealand
Geographical Society, Auckland).
COHEN, R. [1978]
'Ethnicity: problems and focus in anthropology', Annual
Review of Anthropology, 7, 379-403.
CONNOR, W. [1973]
'The politics of ethnonationalism', Journal of International
Affairs, 27, 1-21.
DICEY, A.V. [1939]
Introduction to the Study of the Law of the Constitution,
Ninth edition, (Macmillan, London).
DOFNY, J. and A. AKIWOWO (eds.) [1980]
National and Ethnic Movements, (Sage Publications, Palo
Alto).
DORION, H. [1968]
Rapport de la Commission d'etude sur l'integrite du
Territoire du Quebec, (Government Printer, Quebec).
ELIADE, M. [1969]
The Sacred and the Profane, (Harper Torchbook, New
York).
EMERSON, R. [1960]
From Empire to Nation: The Rise to Self-Assertion of Asian
and African People, (Beacon Press, Boston).
EMERSON, R. [1971]
'Self-determination', American Journal of International
Law, 65, 459-75.
ENGLISH, P.W. and R.C. MAYFIELD (eds.) [1972]
Man, Space and Environment: Concepts in Human Geography,
(Oxford University Press, New York).
FELD W.J. and G. BOYD [1980]
Comparative Regional Systems, (Pergamon Press, New
York).
FRANCK, T.M. [1976]
'The stealing of the Sahara', American Journal of International
Law, 70, 694-721.
GLASSNER, M.I. and H.J. DE BLIJ [1980]
Systematic Political Geography, (Wiley, New York).
GOTTMANN, J. [1951]
'Geography and international relations', World Politics,
3, 153-73, reprinted in W.A.D. Jackson (ed.), Politics and
Geographic Relations: Readings on the Nature of Political
Geography, (Prentice-Hall, Englewood Cliffs).
GOTTMANN, J. [1973]
The Significance of Territory, (The University Press of
Virginia, Charlottesville).

132

GOTTMANN, J. [1975]

'The evolution of the concept of territory', Social Science Information, 14, nos. 3-4, 29-47.

GOTTMANN, J. (ed.) [1980]

Centre and Periphery: Spatial Vaiation in Politics, (Sage Publications, Beverly Hills).

GREER, S. and P. ORLEANS [1964]

'Political Sociology', in R.E.L. Faris (ed.), Handbook of Modern Sociology, (Rand McNally, Chicago), 808-51.

GRIFFITHS, N. [1973]

The Acadians: Creation of a People, (McGraw-Hill Ryerson Ltd., Toronto).

HALE, G.A. [1982]

'Diaspora versus Ghourba: the territorial restructuring of Palestine', in D.G. Bennett (ed.), Tension Areas of the World, (Park Press, Champaign).

HALL, R.L. (ed.) [1979]

Ethnic Autonomy: Comparative Dynamics, The Americas, Europe and The Developing World, (Pergamon Press, New York).

HARRIS, R.C. [1979]

'Within the fantastic frontier: a geographer's thoughts on Canadian unity', Canadian Geographer, 23, 197-200.

HARRIS, W.W. [1981]

Taking Root: Israeli Settlement in the West Bank, the Golan and Gaza-Sinai, 1967-1980, (Research Studies Press, Chichester).

HARTSHORNE, R. [1950]

'The functional approach in political geography', Annals of the Association of American Geographers, 40, 95-130.

HECHTER, M. [1975]

Internal Colonialism: The Celtic Fringe in British National Development, 1536-1966, (Routledge and Kegan Paul, London).

HERZ, J. [1976]

The Nation-State and the Crisis of World Politics, (David McKay Company, New York).

ISRAELIS REPLY [1973]

The Meaning of Secure Borders, (Tel-Aviv).

JACOBS, J. [1980]

The Question of Separation, (Random House, New York).

JONES, E. [1966]

Human Geography, (Praeger, New York).

JORDAN, T.G. and L. ROWNTREE [1982]

The Human Mosaic: A Thematic Introduction to Cultural Geography, (Oxford University Press, New York).

KLIOT, N. [1982]

'Sense of place lost: the evacuation of Israeli settlements from Sinai', in S. Waterman and N. Kliot (eds.), Contemporary Problems in Political Geography, (University of Haifa, Haifa).

KNIGHT, D.B. [1971]
'Impress of authority and ideology on landscape: A review of some unanswered questions', Tijdschrift voor Economische en Sociale Geografie, 63, 383-87.

KNIGHT, D.B. [1975]
'Maps and constraints or springboards to imaginative thought: future maps of Canada', Bulletin of the Association of Canadian Map Libraries, No. 18, 1-9.

KNIGHT, D.B. [1977]
A Capital for Canada: Conflict and Compromise in the Nineteenth Century, (University of Chicago, Department of Geography, Research Paper #182).

KNIGHT, D.B. [1982a]
'Canada in crisis: the power of regionalisms', in D.G. Bennett (ed.), Tension Areas, 254-97.

KNIGHT, D.B. [1982b]
'Identity and territory: geographical perspectives on nationalism and regionalism', Annals of the Association of American Geographers, 72

KOHN, H. [1944]
The Idea of Nationalism, (Macmillan, New York).

KOHR, L. [1957]
The Breakdown of Nations, (Rinehart, New York).

KUNSTADTER, P. (ed.) [1967]
Southeast Asian Tribes, Minorities and Nations, 3 volumes, (Princeton University Press, Princeton).

LEVESQUE, R. [1978]
La Passion du Quebec, (Editions Quebec/Amerique, Montreal).

LOWENTHAL, D. [1961]
'Geography, Experience and Imagination: towards a geographical epistemology', Annals of the Association of American Geographers, 51, 241-60.

LOWENTHAL, D. and M.J. BOWDEN [1976]
Geographies of the Mind, (Oxford University Press, New York]

McCANN, L.D. (ed.), [1982]
Heartland and Hinterland: A Geography of Canada, (Prentice-Hall, Toronto).

McCRAE, K.D. [1974]
Consociational Democracy: Political Accommodation in Segmented Societies, (McClelland and Stewart, Toronto).

McWHINNEY, E. [1982]
Canada and the Constitution 1979-1982: Patriation and the Charter of Rights, (University of Toronto Press, Toronto).

MERRILL, G. [1968]
'Regionalism and nationalism', in J. Warkentin (ed.), Canada: A Geographical Interpretation, (Methuen, Toronto).

MIKESELL, M.W. [1968]
   'Landscape', in D. Sills (ed.), <u>International Encyclopedia
   of the Social Sciences</u>, (Crowell Collier and Macmillan,
   New York).
MINISTRY OF FOREIGN AFFAIRS [1982]
   <u>The Golan Heights</u>, (Jerusalem).
MORGAN, K.O. [1981]
   <u>Rebirth of a Nation: Wales 1880-1980</u>, (Oxford University
   Press, New York).
MORSE, S.J. [1980]
   'National identity from a social psychological perspective
   - a study of university students in Saskatchewan', <u>Canadian
   Review of Studies in Nationalism</u>, 7, 299-312.
NICHOLSON, N. [1979]
   <u>The Boundaries of the Canadian Confederation</u>, (Carleton
   Library Series, No. 115, Macmillan, Toronto).
NINCIC, D. [1970]
   <u>The Problem of Sovereignty in the Charter and in the
   Practice of the United Nations</u>, (Martinus Nijhoff, The
   Hague).
OFFICE OF NATIVE CLAIMS [1978]
   <u>Native Claims - Policy, Processes and Perspectives</u>, (Minister
   of Supply and Services Canada, Ottawa).
PATRICK, R.A. [1976]
   <u>Political Geography and the Cyprus Conflict : 1963- 1971</u>,
   University of Waterloo, Department of Geography Publication
   Series, No. 4, Waterloo).
POLYVIOU, P.G. [1980]
   <u>Cyprus: Conflict and Negotiation 1960-1980</u>, (Duckworth,
   London).
PRATT, L. and G. STEVENSON (eds.) [1981]
   <u>Western Separatism: The Myths, Realities and Dangers</u>,
   (Hurtig Publishers, Edmonton).
QUEBEC LIBERAL PARTY [1980]
   <u>A New Canadian Federation</u>, (Montreal).
RABY, S. [1974]
   'Aboriginal territorial aspirations in political geography',
   <u>Proceedings of the International Geographical Union Regional
   Conference</u> (New Zealand Geographical Society, Palmerston
   North).
RAY, D.M. [1971]
   <u>Dimensions of Canadian Regionalism</u>, (Geographical Paper
   No. 49, Department of Energy, Mines and Resources,
   Ottawa).
SAARINEN, T.F. [1976]
   <u>Environmental Planning: Perception and Behavior</u>, (Houghton
   Mifflin, Boston).
SCHUMACHER, E.F. [1974]
   <u>Small is Beautiful</u>, (Sphere Books, London).

SHAFER, B.C. [1955]
  Nationalism: Myth and Reality, (Harcourt, Brace and World, New York).
SHAFER, B.C. [1966]
  Nationalism: Interpretations and Interpreters, Third edition, (American Historical Society, Washington, D.C.).
SHAFER, B.C. [1972]
  Faces of Nationalism: New Realities and Old Myths, (Harcourt, Brace, Jovanovich, New York).
SMITH, A.D. [1971]
  Theories of Nationalism, (Duckworth, London).
SMITH, A.D. [1979]
  'Towards a theory of ethnic separatism', Ethnic and Racial Studies, 2.
SMITH, G.E. [1979]
  'Political geography and the theoretical study of the East European nation', Indian Journal of Political Science, 40, 59-83.
THE TASK FORCE ON CANADIAN UNITY [1979]
  Three volumes (Supply and Services, Ottawa).
TUAN, Y-F. [1974]
  Topophilia, (Prentice-Hall, Englewood Cliffs).
TUAN, Y-F. [1979]
  Landscapes of Fear, (University of Minnesota Press, Minneapolis).
WATSON, J.W. [1968]
  Canada: Problems and Prospects, (Longmans, Don Mills).
WHEARE, K.C. [1963]
  Federal Government, Fourth edition, (Oxford University Press, London).
WHEBELL, C.F.J. [1973]
  'A model of territorial separation', Proceedings, Association of American Geographers, 295-98.
WILLIAMS, C.H. [1976]
  'Cultural nationalism in Wales', Canadian Review of Studies in Nationalism, 4, 15-38.
WILLIAMS, C.H. [1980]
  'Ethnic separatism in Western Europe', Tijdschrift voor Economische en Sociale Geografie, 71, 142-58.
WILLIAMS, C.H. [1981]
  'Identity through autonomy: ethnic separatism in Quebec', in A.D. Burnett and P.J. Taylor (eds.), Political Studies, 275-89.
WILLIAMS, C.H. (ed.) [1982]
  National Separatism, (University of Wales Press, Cardiff).
WILSON, G. and M. WILSON [1966]
  The Analysis of Social Change, (Cambridge University Press, Cambridge).

WIRSING, R.G. (ed.) [1981]
    Protection of Ethnic Minorities: Comparative Perspectives,
    (Pergamon Press, New York).

Chapter 12
THE INTERNATIONAL FRONTIER
IN MICROCOSM
— THE SHANKILL-FALLS DIVIDE; BELFAST

Frederick W. Boal and David N. Livingstone

THE IDEA OF THE FRONTIER

The term 'frontier' has become commonplace in Western
literature, applied metaphorically, to such diverse subjects as
'the frontiers of science', 'frontiers of the mind' 'frontiersman',
'research frontiers' and 'political frontiers'. And yet, despite
its widespread currency, it is a word which is curiously ambivalent,
vague and lacking distinctive definition. It is difficult to
decide whether it suggests integration or division, an area of
new opportunity or imminent danger, a forward orientation towards
the unknown or a backward inclination towards security. Perhaps
it is this very openendedness that makes is such a potent
image.

Derived from the Latin root <u>frons</u> (front or forepart) by
way of the later mediaeval Latin term <u>fronteria</u> (frontier or
line of battle) (Juricek 1966), the word 'frontier' was defined
in 1623 by Henry Cockeram as 'the bounds or limits of a country'
(Mood 1948), and was later expanded by Nathan Bailey to 'the
Limits or Borders of a Country or Province'. Three major senses
of the word became prominent both in Europe and America during
the eighteenth and nineteenth centuries before Turner added his
own distinctive contribution. By far the most common usage,
according to the <u>Oxford English Dictionary</u> was that of 'a fortress
on the frontier; a frontier town'. This sense of fortress or
outpost on the territorial limits of a nation, province or country
became so commonplace that Benjamin Franklin could, in 1754,
observe that 'the British Colonies, bordering on the French, are
properly Frontiers of the British Empire', while to Duncan
McArthur, in 1815, 'every house in the territory of Michigan is
a frontier' (quoted in Juricek 1966). Clearly, as these – and
many other instances cited by Juricek – confirm, the idea of
frontier as outpost or stockade was the commonest use of the
term and could be applied at widely differing scales from the
individual home, through community town to province or state.

Closely associated with this fortress definition was that
of frontier as a 'barrier against attack'. According to Thomas

Nairn in 1708, South Carolina was 'a frontier, both against the French and the Spaniards' (quoted in Crane 1956: 93). And as an outgrowth of these two meanings, a third, rather more distinctively American notion, of 'a settler on the frontier; a frontiersman' emerged, although this was by far the least common interpretation.

In contrast, F. J. Turner's frontier (1920) was directed westward away from the centres of society and population. Thus whereas earlier frontier conceptions tended to imply a static zone between different peoples and enclosure of the national population and mechanisms for military defence, Turner's frontier was dynamic, processual, abutting unoccupied territory and carrying with it positive images of progress, expansion, novelty and regeneration. Thus, Mood, quoting from Horwill's A Dictionary of Modern American Usage, refers to 'the peculiar Am[erican] meaning of frontier (Mood 1948: 82), and even more recently Hennessy (1978: 7) has pointed out that 'the American usage of the word has none of the restrictive connotations of European frontiers which are seen as 'fortified boundary lines running through dense populations'.

The differences between the older English conception of frontier and the Turnerian and post-Turnerian American version seem to reduce to four polarities. The European concept is restrictive, associated with outer limits and fortifications; Turner's frontier was generative, creating 'a new man' in whom the romantic self-image of the American nation could be invested (Coleman 1966). Secondly, the European model suggests a degree of peripheral introversion, with the backward psychological orientation of the outpost towards the homeland; the American image suggests extroversion, the pioneer frontiersman ever pressing forward to new unexplored regions. Thirdly, the integrative quality of the American experience, typified by Kristof (1959: 273) as providing 'an excellent opportunity for mutual interpenetration' contrasts with the rather more segregative notions of barriers, limits and bounds suggested, for instance, when De Guilleville's 1483 Pylgrymage of the Sowle refers to keeping 'the frontiers of the realm from perille of enemyes' (McDonald 1978). Finally, whereas the American frontier was dynamic in the sense of being processual and indeed social evolutionary, the European connotations, already typified, suggest more static association.

CHARACTERISTICS OF FRONTIER ENVIRONMENTS

Despite constrasts between European and American conceptions of the frontier, several important sets of characteristics common to both types of frontier environment can be derived. The shared features are sparsity and culture contrast; uncertainty, fear and violence; a lack of information, and finally, peripherality.

Sparsity and Culture Contrast

Turner was very clear about what he considered to be the vital ingredient of the frontier: it was the 'hither edge of free land' (1920: 3). This idea of sparsity, first adopted by the original compilers of the U.S. Census, has continued to typify the American usage of the term (See Billington 1950: 3). Similarly, Pierson regarded it as 'an empty land, a surface not only raw but unsettled' (1940: 454), Hennessy (1978: 16) as 'the cutting edge between the settled and unsettled land', and, most recently in his exposition of time frontiers, Melbin writes:

> A frontier is a pattern of sparse settlement
> in space or time, located between a more
> densely settled and a practically empty
> region. Below a certain density of active
> people, a given space-time region is a
> wilderness (1978: 6).

Though these specifications are essentially American, contrasting with the European 'line through dense populations', the notion of sparsity can nevertheless be regarded as essential to identifying a frontier environment. To do this, the concept is to be construed qualitatively as the absence of a culture or a population like that behind the frontier zone. Indeed, in many ways, this is what must have been meant even in the North American context for the presence of an Indian population meant that the West was not totally unsettled – a fact displayed ideologically by Turner himself when he spoke of the frontier as 'the meeting point between savagery and civilisation' (Stamp 1961: 201). Thus, while Turner did not view the frontier 'as a borderland between two ecumenes, between two different types of human societies' (Kristof 1959: 274), his interpretation depended on making the Indian invisible by subsuming both him and his culture under the term 'savagery'.

This juxtaposition of different human groups so as to produce a frontier environment implies the existence of cultural contrasts between the groups. While Mill's definition, at the turn of the century, of frontier as 'the country contiguous between two territories' (Stamp 1961: 201) does not necessarily imply cultural discontinuity, contemporary research on territorial behaviour at different scales suggests at least some degree of perceived dissimilarity (Soja 1971: 34). Indeed, successive definitions of the term in Webster's International Dictionary of the English Language reveal a progressive evolution in meaning from 'a limit, boundary, border or another country' (1806), through 'the part furthest advanced, or that part that fronts an enemy' (1828) to include, as one definition in 1919, 'any line of division between opposites', (Juricek 1966). Such cultural assymetry is particularly evident in the social cleavage between the occupants of a frontier stockade and those outside. Frontier environments are thus marked by some degree of cultural diversity which may take the form of cultural pluralism, integration or segregation, the last being the most likely.[1]

Stress

Historically frontiers have been associated with uncertainty, fear and violence. Thus, in his resume of the qualities of the Roman castra, David Ley (1974: 100) emphasises 'enveloping environment of stress and uncertainty'. There has been a long association of frontier living with violence arising from a variety of sources including external hostility, the nature of a peripheral society and the inapplicability of the legal process to marginal populations. Walter P. Webb, for example, believed that the American West was typefied by lawlessness because 'the law that was applied there was not made for the conditions that existed . . . . It did not fit the needs of the country, and could not be obeyed' (quoted in Melbin 1978: 10). Billington (1950) maintained there was a lack of policemen and that law enforcement agencies were few.

While the extent of frontier violence and its associated mythology in the United States have been seriously questioned (Slotkin 1973; Hollon 1974), the continued vitality of the 'Wild West' myth is in itself an interesting manifestation of contemporary understanding of frontier society in popular culture. The fact is, according to Webb, that what happened on the Great Plains frontier 'was magnified in the press and exaggerated in the imagination, and nothing was more magnified than its unconventionality, its romantic aspects and its lawlessness' (Webb 1931: 502). This suggests that the violence associated in popular consciousness with frontier life may often be the product of imagination or reportorial overkill.

Lack of Information

A lack of informed knowledge about the realities of the external social and physical environment is also typical of a frontier environment. Ley's claim that Roman frontier outposts were characterised by 'a paucity of information concerning the external environment and the control of preconceived images in decision-making' (1974: 100) was replicated in the North American experience. Research on the 'myth' of the great American desert — largely based on the selective experience, both in space and time, of pioneers — and other misconceptions concerning the Great Plains, has shown how scanty was specific knowledge about the frontier zone and how important European preconceptions were in structuring the various images. Indeed, the myth of a western desert, as Lewis (1965) has shown, was itself far from monolithic and various types of desert - physical, demographic and agricultural - were all invoked.

Peripherality

A final characteristic of frontier environments is their peripherality. Physically, politically, militarily and psychologically, frontiersmen are located on the periphery of their societies and, as such, tend to look for social, economic and in some cases military support from their national heartlands. Indeed, the idea of peripherality has been part of the image of frontier society ever since Walker in 1791 spoke of the frontier as the 'utmost verge of any territory' and Webster defined it in terms of 'furthest settlements' (quoted in Mood 1948: 79) (Figure 12.1).

THE SHANKILL FALLS ENVIRONMENT

Sparsity and culture contrast

The spatial juxtaposition of the differing population groups in Belfast in general and in the Shankill-Falls area in particular is a consequence of two sets of population movements. The first located two, at least partially, distinctive peoples in the north of Ireland, while the second brought a portion of each of these groups together in Belfast.

The introduction of ethnic differentiation into the Northern Irish scene is usually associated with the Ulster Plantation in the early seventeenth century (Robinson 1982) when significant numbers of Scots and English were settled on lands confiscated from the Irish. In addition, a strongly based Scottish colony had been estab'ished in the extreme northeastern counties of Antrim and Down (Stewart 1977: 37).

Plantation and colonisation created a predominantly rural population base which preceded the rapid rise of a major industrial complex in the Belfast area in the nineteenth century. While urbanisation sucked in both Roman Catholic and Protestant it did not produce a spatial mingling of the two groups (Jones 1960: 187-192).

The sharp line of separation between the Catholic Falls and Protestant Shankill areas, established in the nineteenth century, has continued to display a remarkable tenacity throughout the present century. Here we focus on a portion of the periphery of Shankill-Falls, South Shankill-Clonard. South Shankill and Clonard interface along the line of a single street (Cupar Street) (Figure 12.2). The pattern of religious distribution in 1911 (Boal 1978), which reveals a sharp line of demarcation along the Cupar Street interface, has remained unchanged (Evans 1944). Similarly, the pattern of population distribution in 1951 (Jones 1960: 196) and at the end of 1967 (Boal 1969: Figure 12.2) continue to provide clear visual evidence of religious segregation across a sharp line of divide at Cupar Street. Since the outbreak of violence in 1969, a series of maps of the distribution of

Figure 12.1: Characteristics of Frontier Environments

POPULATION SPARSITY

CULTURAL CONTRAST

STRESS

OPEN		CLOSED
generative	LACK OF INFORMATION	restrictive
extroversion		introversion
integrative	PERIPHERALITY	segregative
dynamic		static

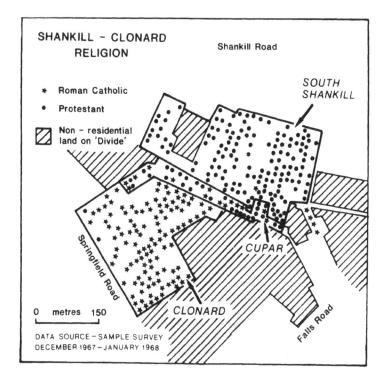

Figure 12.2: Distribution of Protestants and Catholics
South Shankill - Clonard Frontier Zone

143

Figure 12.3: Distribution of Protestants and Catholics and 'Mixed Areas' in Belfast

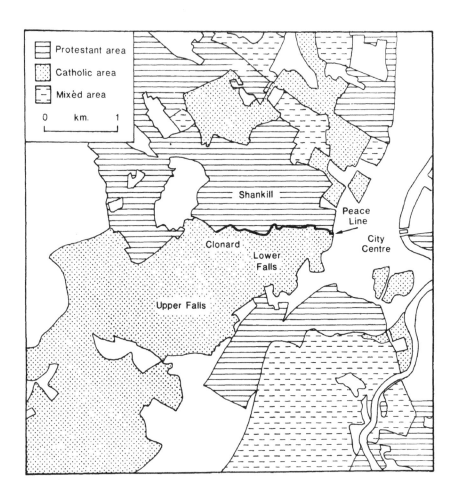

Protestants and Catholics (G.S.G.S. 1969; 1971; 1973; 1975; 1977; 1979) confirms the continuity of segregation in the Shankill-Falls area in general and in the South Shankill-Clonard area in particular, and, in addition, the geographical stability of the boundary line between the two groups (Figure 12.3).

Since language is of major significance in the maintenance of cultural identity, it is important to note that the colonisation process created a language difference in the north of Ireland - the settlers speaking English, the indigenous population Gaelic.[2] Despite this, however, the religious difference was probably crucial for as Stewart (1977) has noted, religion was the barrier which cut the settlers off from the native Irish and placed them permanently in a state of siege. The passage of time led to an elaboration of such cultural differences, as nationalism became a dominant concern in the nineteenth century. Lyons (1978) emphasises the importance of the so-called Gaelic Revival, where Irish language, literature, music and sport were developed or rediscovered as part of the construction of a distinctive Irish identity which was perceived as a Roman Catholic artifact, thus widening the cultural gulf between Protestant and Catholic. Political developments in the late nineteenth and early twentieth centuries indicating a separation of Ireland from Britain were met by a Protestant response that required a maintenance of this British link. Thus, with the conscious reinforcement of cultural distinctiveness on the Catholic side, the Protestants, in turn, defined and emphasised their distinctiveness from the Irish Catholics.

Cultural differences between these groups may seem insignificant when set against the shared array of standard attributes typical of the Western developed world. However, since it is the significance which the two groups in Belfast attach to these often subtle cultural contrasts (rather than any so-called objective measure of difference) that is so important, it is all the more necessary to specify the nature of these differences.

The religious divide is crucial. Despite the overarching Christian frame, there is a radical cleavage between Protestant and Catholic: different churches are attended and the nature of religious observance is distinctive. Further, the assumed uniformity on the Catholic side and diversity on the Protestant, feeds across into the political sphere. The lack of Protestant denominational unity creates the felt need for such a unifying structure as the Orange Order (Harris 1972: 156-157). The religious factor also has an impact on the family; Roman Catholic strictures regarding contraception result in Catholic families significantly larger, on the average, than Protestants (Compton 1981).

In education, separate school and teacher-training systems have emerged as part of the social complex. These reduce peer group interaction to a low level and are instruments for the inculcation of differing views of Irish history. It is claimed that Roman Catholic schools and teacher-training colleges sustain the 'Catholic ethos'. Since sport is an important part of school life, the fact that the separate school systems emphasise a range of separate sporting activities (for instance Gaelic football, hurling and Irish handball in Catholic schools, cricket, rugby football and field hockey in Protestant schools) further maintains cultural differences.

Schools, together with home and local social environment, also transmit culture through language. The most obvious form occurs where Irish is taught in Catholic schools but not in Protestant. More subtly, it occurs in the pronunciation of words (L. Milroy 1980) or individual letters (J. Milroy 1981: 43-44).

The awareness of cultural differences also emerges when members of the two groups are asked to describe themselves. Roman Catholics in Belfast have labelled themselves 'Irish' and 'Nationalistic', Protestants 'British' and 'Loyalist', all these terms being suggestive of a whole array of cultural and political characteristics (O'Donnell 1977). At the same time, the 'ordinary people' has been employed by each group to describe the other, thus serving as a reminder of a certain common ground between the two groups. However, as has already been pointed out, it is this common ground that provides the backcloth to the sharpened salience of those differences that do exist.

Stress and Violence

The presence of stress is a further characteristic of frontier environments, and is usually derived from a sense of insecurity or externally based threat. The general conflict between Catholic and Protestant in Belfast has manifested itself violently in 1835, 1843, 1857, 1864, 1872, 1880, 1884, 1886, 1898, 1920-1922, 1935, 1964 and since 1969. Such outbursts are likely to create particular stress in a situation where the conflicting groups are in close proximity whether or not specific incidents occur in the immediate locality. In addition to this generalised stress, the area of the Divide itself has experienced a long history of violence, for instance, in incidents recorded as the 'Battle of Kashmir Road' (1921), and the major outbreak which occurred in 1969 (Tribunal of Inquiry 1972; Boal 1978). The stressful environment, largely a product of the perceived presence of threat[3], is created and maintained not only by the folk memory of violent events but also by the regular visual displays of the 'other side' such as flags, bunting and kerb painting.

Uncertainty and Lack of Information

We have suggested that uncertainty and a lack of information about conditions beyond the stockade are also characteristic of frontier environments. In Belfast the general air of uncertainty is illustrated by the following:

146

'Belfast today is often compared with London
during the blitz. But at least the Londoners
were united against an identifiable foreign
foe and not exposed to the furtive exploits
of members of their own society. It is the
unpredictability rather than the frequency
of Belfast's hazards that make them so
nerve-wracking. . . .' (Murphy 1979: 134).

The report on the civil disturbances in Belfast in 1969 (Tribunal
of Enquiry 1972) is replete with references to rumours - 'early
in the evening of 14 August there had been rumours among the
Catholic residents of Conway Street that the Protestants planned
to attack them that night' (page 150); 'she was disturbed and
nervous at a rumour that Protestants in Cupar Street were going
to be burned out' (page 196). Lack of information about those
beyond one's own stockade contributes to the prevailing atmosphere
of uncertainty. In the mid-1970s an eighteen year-old boy in
Clonard (the part of the Falls area abutting the Shankill) is
recorded as saying: 'when the troubles are over, I'd love to
visit the Shankill, just to see how they live' (Milroy 1976:
25). Four other youngsters in the room at the time agreed. As
Milroy put it, 'they had not managed the two-minute walk
either'.

Peripherality

A final frontier characteristic is that of peripherality.
Later, we suggest that a situation of multiple peripherality
exists in the Shankill-Falls area of west Belfast. Here it will
suffice to note that the Clonard area is peripheral to the much
more extensive Roman Catholic Falls sector, while the Protestant
area in the vicinity of Cupar Street (South Shankill) is peripheral
to the much larger Shankill.

BEHAVIOURAL RESPONSES TO THE FRONTIER ENVIRONMENT

Given the cultural contrast, stress, uncertainty, violence
and peripherality of the Shankill-Falls Divide, several distinct
responses to the prevailing socio-geographical conditions are
to be expected. The reaction of frontier-dwellers to the
circumstances of their environment centres on a process of
'stockading' - the behavioural complex of responses to frontier
conditions. Subsumed under this process are two distinctive but
highly interdependent sets of mechanisms: those devoted to
securing the periphery and those to consolidating the core.
During the hundred years that the Shankill-Falls frontier
zone has been in existence, it has fluctuated along the open-
closed continuum. For our purposes the examination of responses

147

to the frontier environment can be divided into two periods –
before August 1969, when it displayed a small degree of openness,
and after that date when it became almost completely closed.

Before August 1969

Patterns of segregated activity are one of the most basic
human responses to environments of threat and uncertainty;
individuals avoid venturing out beyond their own territory into
space occupied by those perceived as potential or actual enemies
while, at the same time, inward orientation of activity strengthens
the territorial core.' In the pre-1969 period, social visiting
between the Protestant and Catholic sections of the frontier
zone and movement for shopping or to bus stops were subject to
a high level of ethnic territorial constraint (Boal 1969). The
sense of insecurity and threat beyond one's own territorial
stockade is suggested by the following:

> 'The Border in those days was the New Lodge
> Road. My purpose in making the odd safari
> into Comanche territory' . . . (McAutry 1981:
> 87).

> ' . . . the Mountain was inaccessible because
> to reach it we had to cross territory held
> by the Mickeys [Roman Catholics]. Being
> children of the staunch Protestant quarter,
> to go near the Catholic idolators was more
> than we dared, for fear of having one of our
> members cut off' (Harbinson 1960: 16).

It must be emphasised, however, that during this period the
frontier was not completely closed. There was some movement
between territories for shopping and there was also a narrow
belt of Protestant and Roman Catholic residential mixing in Cupar
Street where even a little interethnic social visiting took
place.

Cores within the frontier zone were maintained by intense
levels of interaction amongst the residents, particularly in
terms of social visiting where well-developed closely-knit
networks could be observed, founded on the twin bases of common
class and common ethnicity consolidated by long-term residence
and by short-distance movement patterns where residential mobility
had occurred (Boal 1969; Weiner 1976; Milroy 1980). Core
characteristics were reinforced by churches and schools functioning
as ethnic institutions, transmitting the cultural norms and
traditional beliefs of the group. Ethnic purity within each of
the separate segments was ensured by religious and socio- political
taboos that favoured endogamy.

The use of symbols is a very important feature of behaviour both in the core and peripheral dimensions of the frontier. As Rowntree and Conkey have written:

> 'Border messages are directed at outsiders
> and are aggressive reminders of a cultural
> boundary. Core or home based messages are
> inwardly directed information that nurture
> and promote group loyalty and identity'
> (1980: 462).

Perhaps the most striking symbolic expression of group territorial identity takes the form of wall graffiti which may be internally or externally directed (Ley and Cybriwsky 1974). While graffiti symbolism on the periphery between the Protestant and Catholic segments of the frontier zone was outwardly directed, symbolism in the core was designed to reinforce group loyalties, not least by threatening deviants or potential traitors. The meaning of some graffiti will be obvious to all; others will have meaning only to a limited group. The level of graffiti in the pre- August 1969 period was relatively muted compared to that which appeared subsequently in the high conflict 1970s and early 1980s. In addition, core reinforcement was also undertaken by the widespread display of appropriate flags and bunting and by the painting of kerbstones in appropriate ethnic livery (red, white and blue in Protestant areas; green, white and orange in Roman Catholic.

Physical stockading was not evident in the pre-August 1969 period - the various streets (Cupar Street, Kane Street, Kashmir Road) leading across the frontier zone from the Protestant to the Catholic segments were unobstructed. Nevertheless, a sharp boundary within the frontier zone was perceived to exist by the residents of both cores, lying along Cupar Street (Boal 1969: 47-49).

August 1969 - December 1981

August 1969 was a particularly traumatic period in Northern Ireland. Rioting occurred in several towns and a number of people were killed and injured (Tribunal of Inquiry 1972). At the same time, within the South Shankill-Clonard frontier zone an outburst of physical conflict occurred triggered by events and rumours of events that were geographically remote from the zone itself. This, in turn, led to a sharpening of behavioural responses to the now even more threatening frontier environment.

On the peripheries of the two frontier segments highly distinctive behaviour patterns emerged. Territorial purification occurred where Roman Catholics and Protestants living outside their respective territorial cores retreated into them for safety[4], or were driven out of their ethnically isolated residences as part of a process of removal of possible 'enemy agents'. The narrow zone of residential mixing along Cupar Street disappeared; the buffer zone within the frontier had collapsed to a line. At the same time as these residential movements were taking place, the limited social interaction between the two segments of the frontier zone almost completely ceased, with the exception of a very small amount of shopping movement undertaken entirely by females, who appeared to have a degree of immunity not granted to males.

Physical stockades, hastily erected by both Roman Catholics and Protestants from whatever materials were locally available replaced the invisible barriers of the earlier period. These included hi-jacked vehicles, building materials, rubble and general junk. The Catholic barricades were the more substantial at this time because most of the forays were from Protestant to Catholic territory. Furthermore, since the authorities seemed to be unable to provide adequate protection, the local communities undertook the organisation of their own defence through the institution of vigilante patrols (Tribunal of Inquiry 1972: 196). Indeed the notion of locally based defence has even been absorbed into the mythology of the frontier community. The Official report (Tribunal of Inquiry 1972) of the shooting of a fifteen year-old boy suggests that he was just a unfortunate caught in the wrong place at the wrong time. However, a wall plaque erected subsequently refers to him as a <u>Fian</u> (warrior) killed 'defending the people of Bombay Street'. True or not, this forms a powerful symbolic device combining the perceived need for peripheral defence with reinforcement of core loyalties. Finally, the invigorated peripheral behaviour was further illustrated in an increase in outwardly directed graffiti and in a competetitive overlaying of graffiti at the few locations to which both Roman Catholics and Protestants had access (Barr 1982; Murtagh 1982; Smyth 1982).

Core reinforcement behaviour was also intensified. Symbols (particularly graffiti, but also flag displays) were used to emphasise and to encourage core loyalties. Wall messages threatened retaliation for any actions that might undermine the security of the core - 'Informers Beware'. Physical retaliation for such deviance or suspected deviance was meted out in ritual punishments or executions - 'tarring-and-feathering', 'knee-jobs', or 'head-jobs' in the singularly brutal jargon of the paramilitaries (Figure 12.4).

Physical stockading, the almost complete cessation of movement across the Divide, and an increased sense of security led to a number of developments that, by increasing self-sufficiency, reduced external dependence. The organisation of vigilantes required regular meetings:

> ' . . . they began to meet together regularly
> for and on behalf of their neighbours and,
> by virtue of understanding this task, began
> to learn the skills of organising and relating

their activities to other organisations. .
. . Some moved towards the formation of
social clubs which would provide an opportunity
for the continuation of the new local spirit
of camaraderie and which would allow the
members to find their recreation within their
own communities' (Griffiths 1978: 170).

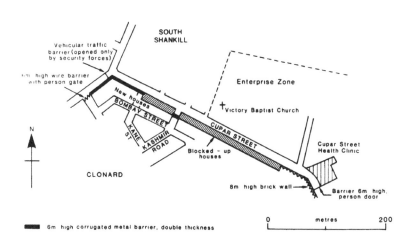

Figure 12.4: Physical Stockading in South Shankill-Clonard
1981

From this there sprang a proliferation of community associations,
youth clubs and advice centres, all organised on a local self-
help basis. Thus not only was the provision of security localised,
but so, too, were concerns about recreational and welfare
provision. Indeed, systems of locally operated public transport
emerged (the 'Black Taxis'), local newspapers and newsheets
enjoyed a wide circulation and, for a time, local (and illegal)
radio stations operated to provide news and morale-boosting
references to the 'brave men on the barricades'. Organisation
within the frontier cores also applied itself towards further
goals related to core reinforcement. On the Shankill side a 'Save
the Shankill Campaign' was mounted to resist urban renewal
policies that were perceived as having the effect of partly
depopulating the area. On the Falls side local activists also
pressed authority on housing matters.

still part of the U.K. state. Thus a range of services continued to be supplied to both the Roman Catholic and the Protestant frontier segments - water, sewage, electricity, telephone, hospital services, garbage disposal and so on. However the most dramatic manifestation of the U.K. state in the frontier zone occurred with the intervention, on August 15th 1969, of the British Army as a peace-keeping force between the Protestant and Catholic territorial factions. Some of the initial difficulties encountered by the Army in regard to the frontier geography were fairly rapidly resolved, thanks to local guidance. By early September 1969 it had been decided that the locally built barricades should be removed and replaced by an official so-called 'Peace Line' erected by the Army. The 'Peace Line' is still in existence - indeed, it has been made more substantial and higher with time (Figures 12.3 and 12.4) despite the insistence of the Guardian newspaper, dated September 10 1969, that 'It should prove useful as a temporary expedient but it must be temporary'. Decisions on the precise location of the Peace Line demonstrate, in a most dramatic fashion, concern for geographical minutiae, involving discussions between locals and the highest military authorities.

INCOMPATIBLE TERRITORIAL NESTING AND MULTIPLE PERIPHERALITY

The prevailing socio-geographical conditions along the Shankill-Falls frontier cannot be understood solely by reference to circumstances endemic to the zone itself. In a stimulus response fashion, the local scene is both directly and indirectly influenced by an enveloping cultural- political environment. The highly dynamic nature of the social system means that events in one part have profound ramifications elsewhere, such as increased tension, intimidation and violence. This implies that understanding the territorial behaviour of the two groups in question requires a contextual rather than a solely internal analysis. It is suggested that the two ideas of 'incompatible territorial nesting' and 'multiple peripherality' are needed as additional dimensions of the frontier model as applied to the Shankill-Falls area.

Conventionally the concept of neighbourhood territoriality has been employed in the analysis of the subcultural behaviour of different groups within cities. Despite the manifestations of territorial behaviour which highlight differences between particular human groups, this behaviour has generally taken place within an overarching political system which, in large measure, enjoys a high degree of consensus.

A major raison d'etre of a political system, according to Soja (1971: 33), is to structure its internal domain administratively into an integrated set of territorial compartments. Where this occurs, neighbourhood territories within cities can be said to

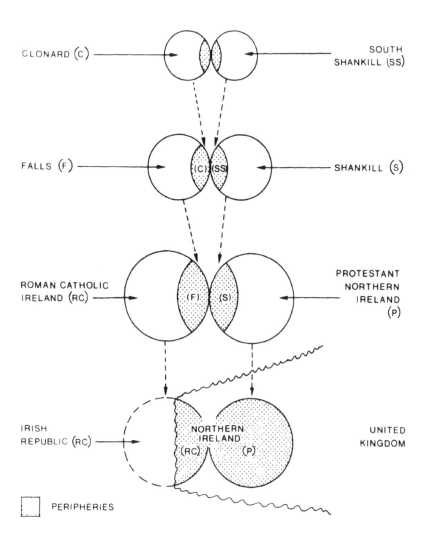

Figure 12.5: Multiple Peripherality and the South
Shankill - Clonard Area, 1981

CLONARD (C)

SOUTH
SHANKILL (SS)

FALLS (F)

(C)(SS)

SHANKILL (S)

ROMAN CATHOLIC
IRELAND (RC)

(F)   (S)

PROTESTANT
NORTHERN
IRELAND
(P)

IRISH
REPUBLIC (RC)

NORTHERN
IRELAND
(RC)   (P)

UNITED
KINGDOM

PERIPHERIES

nest compatibly with each other within the larger polities of city and state. In coming to grips with the behavioural patterns in the Shankill-Falls frontier, however, it must be recognised that the two territories forming the frontier zone do not nest compatibly within the same state - they are parts of two apparently incompatible nationalisms, the Irish on the one hand and the Northern Irish version of the British on the other (Boal, Murray and Poole 1976; Boal 1980). Thus the orientation of the Protestant Shankill is 'Ulster and British' while the Roman Catholic Clonard is 'Celtic Irish' in inclination. This suggests, therefore, that the local frontier in the vicinity of Cupar Street is not just an expression of urban ethnic differences, but is a microcosm of national division.

When this international dimension is recognised, the concept of multiple peripherality also assumes prominence. Clonard and South Shankill (Figure 12.5), with their own local peripheries, can, in turn, be viewed at a larger scale as peripheries of the wider Falls and Shankill areas of West Belfast. The Falls and Shankill, at an even larger scale, can be viewed as peripheries of their respective national cores - these being Roman Catholic Ireland for the Falls and Protestant Northern Ireland for the Shankill. From a somewhat different perspective, Northern Ireland as a whole (with both its Protestant and Roman Catholic components) can be seen as a periphery of the United Kingdom state. Multiple peripherality for both the South Shankill and Clonard segments of the Shankill-Falls frontier zone creates precisely the same situation, at the local level in west Belfast, as that suggested perceptively by Seamus Heaney (1975: 60) for the Roman Catholic community in Northern Ireland as a whole, where they are 'besieged within the siege'). Thus, the frontier residents on both sides of the Peace Line experience multiple peripherality - they are part of the periphery of their more secure ethnic heartlands and they are ultimately on the periphery of two systems of nationhood. Their experience of frontier conditions of stress, uncertainty and deep cultural division arises from the fact that the Shankill-Falls frontier zone is a particularly potent local manifestation of broader political forces at work within the state and beyond.

CONCLUSION

In our analysis of the Shankill-Falls conflict zone in Belfast, the concept of the frontier, as interpreted by us, has proved to be a useful heuristic device. Our analysis indicates that the application of the frontier concept to the study of political and spatial phenomena should not be restricted to one particular scale. In addition, our use of the frontier model leads us to expect that similar applications elsewhere in the world will have considerable potential.

FOOTNOTES

1. According to Lattimore (1962) 'Frontiers are of social, not geographic origin. Only after a concept of a frontier exists can it be attached by the community that has conceived it to a geographical configuration. The consequences of belonging to a group, a group that includes certain people and excludes others, must precede the conscious claim for that group of the right to live or move within a particular territory'.

2. Recently Buchanan (1982) has pointed out that some of the settlers who came over from Scotland were themselves Gaelic speakers.

3. As part of a repertory grid analysis, a random sample of Protestants and Catholics in Belfast was asked, among other things, to assess the elements 'Falls' and 'Shankill' in respect to a construct 'dangerous-safe'. These two elements had to be placed on the construct together with ten other elements. A low score on a scale one to twelve, indicated than an element had been construed 'dangerous'. 'Falls' scored 4.5 for Catholics, 2.0 for Protestants – very dangerous for Protestants and only moderately safe for Catholics. 'Shankill' scored 4.7 for Protestants, 2.3 for Catholics, presenting an almost exact reversal of the pattern of scores for 'Falls'.Thus there was judged to be general danger in the Shankill-Falls area, most marked in the part belong to the other ethnic group.

4. The pressures for relocation are dramatised by recent requests by some Protestants for the exhumation of kin from a cemetery in what is viewed as Roman Catholic territory, and their reburial elsewhere. The exhumation requests stem, in part, from acts of desecration. (Belfast Telegraph, 1 October 1981).

REFERENCES

BARR, J. [1982]
   Geographical interrelationships of sectarianism, territoriality and graffiti in Belfast, Unpublished B.A. Dissertation, University of Oxford.
BILLINGTON, R.A. [1950]
   Westward Expansion, (Macmillan, New York).
BOAL, F.W. [1969]
   'Territoriality on the Shankill-Falls Divide, Belfast', Irish Geography, 6, 30-50.

BOAL, F.W. [1978]
    'Territoriality on the Shankill-Falls Divide, Belfast: The
    perspective from 1976', in D.A. Lanegran and R. Palm (eds.),
    An Invitation to Geography, 2nd edition, (McGraw-Hill, New
    York), 58-77.
BOAL, F.W. [1980]
    'Two nations in Ireland', Antipode, 12(1), 38-44.
ᴾOAL, F.W., R.C. MURRAY and M.A. POOLE [1976]
    'The urban encapsulation of a national conflict' in S.E.
    Clarke and J.L. Obler (eds.), Urban Ethnic Conflict: A
    Comparative Perspective, Comparative Urban Studies Monograph
    #3, Institute for Research in Social Sciences, (University
    of North Carolina, Chapel Hill), 77-131.
BUCHANAN, R.H. [1982]
    'The planter and the Gael: Cultural dimensions of the
    Northern Ireland problem', in F.W. Boal and J.N.H. Douglas
    (eds.), Integration and Division; Geographical Perspectives
    on the Northern Ireland Problem, (Academic Press, London),
    149-73.
COLEMAN, W. [1966]
    'Science and symbol in the Turner Frontier Hypothesis',
    American Historical Review, 72, 22-49.
COMPTON, P.A. [1981]
    'Review of population trends in Northern Ireland, 1971-
    78', in P.A. Compton (ed.), The Contemporary Population
    of Northern Ireland and Population-related Issues, (Institute
    of Irish Studies, Belfast), 7-21.
CRANE, V.W. [1956]
    The Southern Frontier, (University of Michigan Press, Ann
    Arbor).
EVANS, E.E. [1944]
    'Belfast: The site and city', Ulster Journal of Archaeology,
    Third series, 7, 5-30.
GRIFFITHS, H. [1978]
    'Community reaction and voluntary involvement', in J. Darby
    and A. Williamson (eds.), Violence and the Social Services
    in Northern Ireland, (Heinemann, London), 165-94.
G[EOGRAPHICAL] S[ECTION] G[ENERAL] S[TAFF] - BELFAST
    Religious Areas, 1:15000 (1969), 1:10000 (1971, 1973, 1975,
    1977, 1979).
HARBINSON, R. [1960]
    No Surrender, (Faber and Faber, London).
HARRIS, R. [1972]
    Prejudice and Tolerance in Ulster, (Manchester University
    Press, Manchester).
HEANEY, S. [1969]
    Door Into The Dark, (Faber and Faber, London).
HEANEY, S. [1975]
    North, (Faber and Faber, London).

HENNESSY, A. [1978]

    The Frontier in Latin American History, (Edward Arnold, London).

HOLLON, W.E. [1974]

    Frontier Violence - Another Look, (Oxford University Press, New York).

JONES, E. [1960]

    A Social Geography of Belfast, (Oxford University Press, London).

JURICEK, J.T. [1966]

    'American usage of the word 'frontier' from Colonial times to Frederick Jackson Turner', Proceedings of the American Philosophical Society, 110, 10-34.

KRISTOF, L.K.D. [1959]

    'The nature of boundaries and frontiers', Annals of the Association of American Geographers, 49, 269-82.

LATTIMORE, O. [1962]

    Studies in Frontier History. Collected Papers 1928- 1958, London. (Oxford University Press, London).

LEWIS, G.M. [1965]

    'Three centuries of desert concepts in the Cis-Rocky Mountain West', Journal of the West, 4, 457-68.

LEY, D. [1974]

    The Black Inner City as Frontier Outpost, Washington D.C. (Association of American Geographers, Monograph #7, Washington D.C.).

LEY D. and R. CYBRIWSKY [1974]

    'Urban graffiti as territorial markers', Annals of the Association of American Geographers, 64, 491-505.

LYONS, F.S.L. [1978]

    The Burden of our History, The W.B. Rankin Memorial Lecture, Queen's University, Belfast.

McAUTRY, S. [1981]

    Sam McAutry's Belfast, (Ward River Press, Dublin).

McDONALD, P. [1978]

    'The world of words: Frontiers', Area, 10, 173-74.

MELBIN, M. [1978]

    'Night as frontier', American Sociological Review, 43, 3-22.

MILROY, L. [1976]

    'Phonological correlates to community structure in Belfast', Belfast Working Papers in Language and Linguistics, 1, 1-44.

MILROY, L. [1980]

    Language and Social Networks, (Basil Blackwell, Oxford).

MILROY, J. [1981]

    Regional Accents of English: Belfast, (Blackstaff Press, Belfast).

MURPHY, D. [1979]
  A Place Apart, (Penguin Books, Harmondsworth).
MOOD, F. [1948]
  'Notes on the history of the word frontier', Agricultural
  History, 22, 78-83.
MURTAGH, B. [1982]
  Graffiti in West Belfast. Unpublished B.A. Dissertation,
  Queen's University, Belfast.
O'DONNELL, E.E. [1977]
  Northern Irish Stereotypes, (College of Industrial Relations,
  Dublin).
PIERSON, G.W. [1940]
  'The frontier and frontiersman of Turner's essay',
  Pennsylvania Magazine of History and Biography, 64, 449-
  78.
ROBINSON, P. [1982]
  'Plantation and colonization: The historical background',
  in F.W. Boal and J.N.H. Douglas (eds.), Integration and
  Division, 19-47.
ROWNTREE, L.B. and M.W. CONKEY [1980]
  'Symbolism and the cultural landscape', Annals of the
  Association of American Geographers, 70, 459-74.
SLOTKIN, R. [1973]
  Regeneration Through Violence: the Mythology of the American
  Frontier, 1600-1860, (Columbia University Press, New
  York).
SMYTH, A.J. [1982]
  Graffiti and Territoriality in East Belfast. Unpublished
  B.A. Dissertation, Queen's University, Belfast.
SOJA, E.W. [1971]
  The Political Organization of Space, (Association of
  American Geographers, Resource Paper #8, Washington
  D.C.).
STAMP, L.D. [1961]
  A Glossary of Geographical Terms, (Longmans, London).
STEWART, A.T.Q. [1977]
  The Narrow Ground: Aspects of Ulster 1609-1969, (Faber and
  Faber, London).
TRIBUNAL OF INQUIRY [1972]
  Violence and Civil Disturbances in Northern Ireland in
  1969, (Her Majesty's Stationery Office, Belfast).
TURNER, F.J. [1920]
  The Frontier in American History, (H. Holt, New York).
WEBB, W.P. [1931]
  The Great Plains, (Grosset and Dunlap, New York).
WEINER, R. [1976]
  The Rape and Plunder of the Shankill, (Notaems Press,
  Belfast).

Chapter 13
DISSIMILARITIES IN THE EVOLUTION OF FRONTIER CHARACTERISTICS ALONG BOUNDARIES OF DIFFERING POLITICAL AND CULTURAL REGIONS

Moshe Brawer

The imposition of a newly formed international boundary on an environment within which or in the vicinity of which such a political-geographical feature had not existed in the recent past will result in the initiation of 'frontierisation' processes along both sides of such a boundary.[1] A short time after such a boundary is born and assumes effective functions (in some cases within a few months) some characteristics of frontier areas will develop within a narrow zone adjoining this boundary. The nature and pace of these developments will depend on three main groups of factors: 1) The geographical (physical and human) properties of the environment through which the newly formed boundary runs. 2) The differences in character, quality, policy and practices of the administration (government) of the states separated by this boundary. 3) The actual functions of the boundary. These functions generally reflect the state of relations between the governments of the countries concerned and often also differences in their political, economic and cultural character (Prescott 196 ; Solch 1923; Wilkinson 1955; House 1959; Minghi 1963; Boal and Livingstone, this volume).

Mass communications media have repeatedly given much prominence to reports on the effects, on the population and landscape, of the imposition and later the sealing of the boundary between such states as North and South Korea or between East and West Germany. Nevertheless, even these extreme cases of environmental trasnformation due to 'frontierisation' have not, so far, been subject to comprehensive geographical studies, probably due to the military and political sensitivity of the authorities to any research activities in these areas (see Schwind 1950; 1962; Verhasselt 1966; Framke 1968).

Boundary changes in Israel and the Israeli-occupied territories in the last 35 years have been closely followed by geographers and are well-documented (maps, air-photos, statistical data and scientific surveys). They provide a unique opportunity for detailed geographical studies on effects of newly-formed boundaries. Here, the new boundaries cut across densely inhabited rural areas, desert areas inhabited by nomads and a city divided, Jerusalem. The imposition of a totally sealed new boundary led to rapid frontierisation processes over a narrow strip of territory on both sides. After functioning as a barrier of extreme separation for nearly two decades, it abruptly lost its previous character and became an internal administrative boundary, setting into motion rapid 'de-frontierisation' processes in the frontier zone.

This study is devoted to the section of the Israel- Jordan boundary imposed in 1948-49 on central Palestine, a region known as Judaea and Samaria or 'The West Bank'. Legally, this boundary is only an armistice line which came into being following the first Arab-Israel War in 1949, which resulted in the annexation of most of the hilly region of Central Palestine by the Kingdom of Jordan. Except for a single crossing point in Jerusalem, open only to foreign diplomats and small numbers of foreign tourists, this boundary was sealed, heavily guarded and impassable.

Following the Six-Day War of 1967, Israel occupied The West Bank and as a result, the Green Line became an administrative boundary between Israel occupied territory and Israel proper open to free, uncontrolled movement of people and goods. This change made it possible to examine the dissimilarities in the geographical properties of the frontier zone, on each side of the boundary, which had developed during the period it was sealed and subject to the differing policies, regimes and cultures of Jordan and Israel. It also presented excellent conditions for a continuous study of transformation processes brought about by the elimination of an extreme barrier type international boundary (Figures 13.1 and 13.2).

EXTENT AND CHARACTER OF THE FRONTIER ZONE

Geographical surveys of the rural areas of Judaea and Samaria carried out between 1967 and 1972 indicated clearly that the imposition of the Green Line created a distinct frontier zone, generally 3-5 kms wide, on each side of the boundary. Within this narrow strip, most military exchanges, infiltrations and acts of terrorism took place. Villages and townships within this belt developed local characteristics caused by the proximity of the boundary which were conspicuously different on each side of the line. No such differences had existed in this region prior to the imposition of the Green Line.

An apparent relationship exists between the density of population and types of settlement in the frontier zone and the nature and intensity of the effects of the Green Line. Thus, the division of Jerusalem and its frontierisation have had far-reaching effects on its pattern of growth, socio-economic structure and its urban functions. On the other hand, the Green Line had very little impact on the few nomads in the frontier

Table 13.1

POPULATION ('000s)

(Frontier Zone)

	1947		1967		1980	
	Arabs	Jews	Arabs	Jews	Arabs	Jews
Jordanian	109	0.8	112	0	146	1.8
Israeli*	103	11	64	93	96	112
Jerusalem	60	97	65**	209	108	298

* Excluding the urban area of Jerusalem
** Of whom 2,600 lived in the Israeli part of the city.
There were no Jews in the Jordanian section.

Table 13.2

NUMBER OF SETTLEMENTS
(Excluding Jerusalem and suburbs)

	1947		1967		1980	
	Arabs	Jews	Arabs	Jews	Arabs	Jews
Jordan	108	6	105	0	109	11
Israel	96	26	27	125	27	128

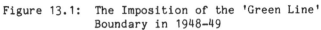
Figure 13.1: The Imposition of the 'Green Line'
Boundary in 1948-49

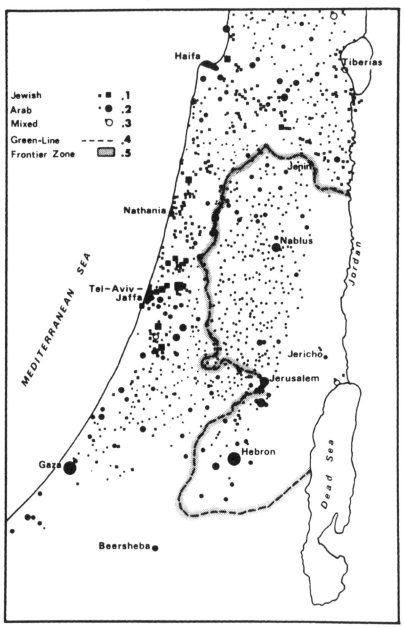

Figure 13.2:   The 'Green Line' in 1979

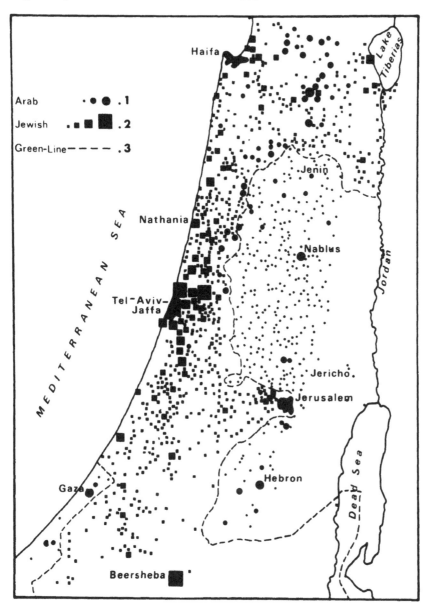

areas of the Judaean Desert. The geographical consequences of the planting of an impassable international boundary through the heart of Jerusalem and the conspicuously different character which each side assumed are subject of another study and are not dealt with here (Cohen 1978).

The dissimilarities in the characteristics of each side of the frontier zone developed primarily due to the basic differences in government policies and practices but also to differences in the nature of the terrain, the ethnic composition of the population, its social and economic structure as well as its level of technology. Jordan's policy, in its section of the frontier zone, mainly expressed itself in attending almost exclusively to military requirements. The infrastructure developed on the Jordanian side of the frontier was intended to serve military needs. It provided some indirect benefits to the local population but these were only minor compensation for the serious hardships which proximity to the new boundary produced. The government was indifferent to the difficulties and requirements of the inhabitants, especially the rural population. The impact of the Green Line on the frontier population, on the Jordanian side, was left to run its 'natural course' with hardly any intervention on the part of the authorities to offset any of the negative effects (Bar-Lavie 1981; Lifschitz 1970). The inhabitants of the frontier zone found themselves a short time after the imposition of the Green Line at a serious disadvantage as compared with the population of other parts of the West Bank. Many of them became destitute, dependent on aid from international relief organisations and/or financial support by relatives who had migrated to other countries, mainly Kuwait and Saudi Arabia. They were also cut off from markets and urban centres which provided them with vital services. It took several years before these services were restored to most villages in the frontier zone.

ISRAEL'S FRONTIER ZONE DEVELOPMENT POLICY

Israel's policy, on the other hand, has been to settle and develop its side of the frontier up to the border line. It provided the population of this zone with privileges and advantages to attract new inhabitants and took special care to meet their needs. Fifty-three new villages (mostly for new immigrants), many of them fortified, were established close to the new border between 1949 and 1954 and entrusted with guarding it. Their inhabitants performed military functions in addition to their normal farming activities. Intensive cultivation of arable lands, often by irrigation with water piped over considerable distances, was strongly encouraged and subsidised. National development projects were designed to populate the frontier zone and exploit its natural resources. Arab villages on the Israeli side of the frontier were also drawn into these dynamic programmes and underwent rapid transformation of their physical pattern and socio-economic character. This policy of establishing numerous Jewish villages on the Israeli side of the frontier zone was also intended to turn the Green Line, along most of its length, into an ethnic boundary.

The dissimilarities between the frontier zones on each side of the Green Line reached their climax in 1967, just before the line lost its status as an international boundary. The differences expressed themselves in growth of population and its spatial distribution, type and organisation of settlements, demographic structure and processes, socio-economic structure economic development, infrastructure and services and technological and cultural progress. Insofar as the Arab inhabitants of the frontier are concerned additional dissimilarities evolved, mainly in behaviour, dress, diet and even language. The Arab population in the Israeli part of the frontier zone became much more 'Europeanised' in its way of life than that of the Jordanian side where the traditional Middle Eastern appearance of the villages and their inhabitants hardly changed (Karmon 1977; Brawer 1978; Shmueli and Schnell, in press).

Table 13.1 represents the changes in numbers and ethnic composition of the population and settlements in the newly-formed frontier zone of the West Bank. The 1947 figures reflect the situation before the imposition of the Green Line while those of 1967 represent the climax of the frontierisation process.

DEPOPULATION OF THE WEST BANK FRONTIER ZONE

The Arab population on the Jordanian side of the frontier zone grew by less than 3 per cent between 1947 and 1967 while the increase in the population of the entire West Bank over the same period was nearly 20 per cent. Further, the average annual natural increase in this zone in the 1950s and 1960s was 2.2 - 2.5 per cent. However, extensive emigration, mainly of young males brought down the actual increase in population in the frontier zone. The depopulation process of the Jordanian side is also reflected in some demographic peculiarities. In 1967, the 15-44 age-group formed 34.1 per cent of the total population of the West Bank. In most villages and townships in the frontier zone this age-group comprised less than 30 per cent, and in some villages was as low as 22-25 per cent. The average for Arab villages on the Israeli side of the frontier zone was 37.2 per cent. Further, in the West Bank as a whole, there were 79.2 males for every 100 females in the 15-44 age-group as against 68.6 males per 100 females in the same age-group in the frontier zone. In some villages, the corresponding figure for males was as low as 51-55. In Arab villages on the Israeli side a small excess of males (100.7) was recorded in the same age-group. Finally, the percentage of people above the age of 65 for the entire West Bank was 6.6 in 1967 and 7.8 in the frontier zone. In Arab villages on the Israeli side of the frontier zone, this age group formed only 3.9 per cent.

While many young males emigrated from the Jordanian side of the frontier zone during 1949-67, the Israeli side of the frontier witnessed a large influx of Jewish settlers, mainly young people. Thus, in addition to the fact that the Jewish population in this zone grew more than eightfold, it included a comparatively high proportion of people of working age. The 15-44 age-group comprised 44.3 per cent of the Jewish population of the Israeli side of the frontier zone as against an average of 42.2 per cent for Israel. There is only a minor discrepancy between males and females in this age-group (98.8 males per 100 females) among Jewish inhabitants of the frontier.

The depopulation process of 1949-67 with its demographic consequences took a sharp turn after 1968 as a result of the fundamental change in the character and functions of the Green Line. Not only did emigration from the former Jordanian side of the frontier zone drop to insignificant dimensions but a growing number of workers from inside the West Bank and from the Gaza Strip settled in this frontier area. Numerous employment opportunities in the densely inhabited Israeli coastal plain, with its rapid economic growth, gave the Green Line frontier zone advantages which attracted many new inhabitants. Between 1968 and 1980, the Arab population of the West Bank grew by an annual average of 3 per cent. This reversal in the population trend is one of the more conspicuous results of turning the Green Line from a boundary of total separation to one of intensive interaction.

On the other hand, with the extension of Israeli military rule to the West Bank, the Israeli side of the Green Line frontier zone lost its political and military significance and its special privileges. It ceased to hold a preferential status as far as development benefits were concerned. Hardly any new Jewish villages were established in this zone after 1968. The rate of increase of the Jewish population slowed down considerably and was mostly confined to urban expansion. The only exception to this is Jerusalem where strenuous efforts have been made to develop and expand the city since 1967.

DIFFERENCES IN PATTERN AND SOCIO-ECONOMIC STRUCTURE

All the Arab villages in what was to become a frontier zone in 1949 were highly clustered with narrow tortuous lanes and walled courts. Socially, the inhabitants of each village were divided into a number of clans, each of which was almost completely confined to its own section of the built-up area so that the social structure of the village was clearly reflected in its pattern. Initial trends towards a more dispersed pattern, associated with a break-up of the clan segregation, were noticeable the latter years of the British Mandate (Kendall 1949).

The establishment of the Green Line boundary had an impact on the subsequent developments in these characteristics of Arab

villages in the newly created frontier areas. On the Jordanian
side, changes in pattern and social structure over the period
1949-67 were slow and very limited. The transformation trends,
which had begun before the birth of the Green Line, came to a
halt in many of the frontier villages. Thus, the West Bank side
of the frontier zone remained almost unchanged and was the area
least affected by the few changes that actually took place on
the West Bank.

During the same period, the Arab villages on the Israeli
side of the frontier zone underwent a rapid and far-reaching
transformation under the influence of development projects and
the general economic growth of the country as a whole. Very
little of the traditional clustered pattern has remained in these
villages which have assumed a new form with modern houses built
along paved streets. The segregation into quarters based on
clan affiliation disappeared simultaneously with a gradual fall
in the weight of the clan in the social and economic structure
(Cohen 1965).

The Jewish villages on the Israeli side are nearly all
cooperative villages (moshav) or communal settlements (kibbutz).
Their physical pattern and its expansion and development are
closely associated with efficient marketing and supply cooperatives
which serve nearly the entire Jewish rural population in Israel.
Some of these frontier villages have industries in addition to
their agricultural activities.

Differences in standards of living

The disparity in the standard of development between the
two sides of the frontier is well represented by several facts.
In 1967, while all villages (Arab and Jewish) on the Israeli
side were fully supplied with modern water and electricity
networks, only 11 of the 105 villages in the West Bank frontier
zone had a modern water supply system (which served only part
of their inhabitants), and only 19 villages had an electricity
supply. None of the villages on both sides of the Green Line had
electricity or running water before 1948. In 1967, only 5.3 per
cent of the dwellings on the West Bank side had running water
and only 6.1 per cent had electricity as against 73.3 per cent
and 66.4 per cent, respectively, in Arab villages on the Israeli
side. This expressed itself in two distinctly different types
of rooftop landscapes. On the Israeli side of the frontier zone,
most roofs carried solar heaters and many had TV antennae,
features absent on the Jordanian side.

Commercial services in frontier villages on the West Bank
were few and consisted mainly of small low standard groceries.
Roving pedlars supplied clothing, household goods and some other
commodities. There were no financial and commercial services
in any of these villages. These services were developed to a
much higher level in Arab villages on the Israeli side. Nine

167

villages had banks, insurance agencies and various commercial enterprises. Nearly all other villages had, in addition to groceries, a variety of other shops.

The differences between both sides of the frontier zone were also expressed in the pattern of employment and sources of maintenance. On the West Bank side in 1967, 78 per cent of the villagers in the frontier zone received their earnings from agriculture. In this regard, little had changed significantly since the Green Line came into being. Of the households in these villages, 46.2 per cent depended on financial assistance from international relief organisations, Arab aid funds and/or relatives (Central Bureau of Statistics 1968). Unemployment, which was high in the first years after the imposition of the Green Line, receded gradually, as a growing number of young people emigrated to the East Bank (Jordan), to the Persian Gulf region and to other countries.

Substantial discrepancies in standards of agricultural production had also developed between both sides of the frontier, the result of differences in the pace and extent of introduction of modern techniques and equipment, use of fertilisers and irrigation. Table 13.3 provides a representative comparison of average agricultural production in the Arab villages of the frontier zone on both sides of the Green Line.

TABLE 13.3

AVERAGE YIELDS 1966-67
(tons per hectare)

	West Bank Side	Israeli Side
Wheat	0.9	1.5
Sesame	0.3	0.52
Tomatoes	8.4	33.2
Cucumbers	6.5	18.3
Grapes	2.9	4.1

Source: Authority for Economic Planning (1967)

By 1980 these differences were reduced. Agricultural yields in the West Bank frontier zone have reached and even exceeded those of 1966-67 in the Arab villages on the Israeli side, while the latter have since attained substantially higher yields.

Decline in Agricultural Occupations

Between 1949 and 1967, agriculture gradually lost its position as the most important source of income in the Arab

168

villages on the Israeli side of the frontier zone. A growing number of villagers found jobs in the urban areas of the coastal plain, mainly in industry, the building trade and services. By 1967, only about 15 per cent of the labour force was still engaged in farming. Nearly 70 per cent commuted daily to work outside the frontier zone. In recent years, there has been a continuous fall in the percentage of villagers engaged in agriculture due to the rapid natural growth of the population and the progressive introduction of new models of labour-saving equipment (Shmueli and Schnell, in press). In Jewish frontier villages, where different types of agriculture are practised, including much mixed farming, 25-30 per cent of the labour force is engaged in agriculture. Another 35 per cent is employed in the villages or their proximity in occupations other than agricultural. Some 27 per cent commute to work in urban areas.

The change in the character and functions of the Green Line had little effect on the pattern of employment and sources of income of the Arab and Jewish inhabitants of the Israeli side. It has brought far-reaching changes in occupations and sources of livelihood to most people on the West Bank side of the frontier zone. By 1980, more than 50 per cent of the adult male labour force of the West Bank frontier villages was employed in Israel, mainly in the coastal plain, and commuted daily from their villages across the Green Line to work. In these West Bank frontier villages, the percentage of villagers who still derive most of their income from farming has fallen to less than 20 per cent. In many of these villages, agricultural activities are carried out by women, children, elderly people and, to some extent, by the adult males on Saturdays (Brawer 1980).

Among the many other differences which developed between the Arab villages which were separated by the Green Line is that of the participation of women in the labour force seeking employment outside their villages. Under Moslem tradition, women were not expected to work outside the family household and farm. When employed outside the property of their family, women had to be accompanied by a close male relative to safeguard their honour. Very few women in the Green Line frontier zone were employed outside their home villages before 1948-49 and these consisted of young girls working as teachers or nurses in neighbouring villages or townships. In 1967, there was little change in this situation among the villages on the West Bank side. However, on the Israeli side, several thousand Arab women commuted daily to work in industry, agriculture and services in Jewish enterprises in the coastal plain. In some of these Israeli-Arab frontier villages females formed nearly 20 per cent of the adult population regularly employed outside their home village. Since the occupation of the West Bank by Israel, there has been a slow but gradual increase in the number of women in villages on that side of the Green Line who work as hired labour outside their home villages. Their numbers and percentage among wage earners on the West Bank side of the frontier zone is still much lower than that of the Arab females on the Israeli side. The same applies to the enrolment of females in secondary schools and institutions of higher education.

Linguistic and other differences

As a result of 19 years of Israeli rule between 1949 and 1967, Arabs on the Israeli side of the frontier zone could not only converse in Hebrew, but many Hebrew expressions had penetrated their Arabic. This was in stark contrast to the Arabic of their neighbours on the other side of the frontier which contained no Hebrew words nor expressions connected with the Israeli way of life.

In addition, considerable differences in the rate of literacy developed among the Arab villages on the two sides of the frontier during this period. Before the imposition of the Green Line, it was estimated that the literacy rate was 26 per cent in the wider region under study, a rise from 11.6 per cent in 1931 (Government of Palestine). By 1967, the figure had risen to 33.2 per cent on the Jordanian side and 50.5 per cent on the Israeli side. Amongst the Jewish villages, the equivalent figure was 89 per cent.

Although since 1967 there have been Hebrew influences in the Arabic spoken on the West Bank side of the frontier zone, the level is lower than these influences had been among the Arab villages on the Israeli side prior to 1967. This is mainly because Hebrew is not taught in the Arab schools on the West Bank.

Similarly, the sale of newspapers and other printed material is almost three times higher in the Arab villages on the Israeli side than among those on the West Bank side. On the other hand, because of its proximity to Israel and the higher level of interaction, the Israeli influence is much more strongly felt in the frontier zone than in other parts of the West Bank.

Among other dissimilarities recorded during work on Arab villages in the frontier zone are those relating to household and dietary habits. In almost every case, the results indicated a higher level of modernisation on the Israeli side of the frontier zone. This is partly explained by the higher general standard of living in the villages in Israel and partly because the imposition of the Green Line created new and different patterns of interaction for the populations on the two sides of the Green Line, forcing them into different political territories and bringing them into close proximity with an extreme barrier-type boundary.

RECENT DEVELOPMENTS

Political developments since 1977 have tended to produce a new situation in the West Bank which may lead to yet another turning point in the character and functions of the Green Line with possible changes in the frontier zones. Under the Egypt-Israel peace treaty, the West Bank should become an autonomous region. An adminisration run by its Arab inhabitants should be set up, based on a different pattern of relationships with Israel than that which had existed since 1967. The Green Line, which will probably have to undergo some rectifications will, thus assume a new role, the nature of which will largely depend on the model of self-government to be instituted in the West Bank.

Another new factor is the policy of the current Israeli government, in office since 1977, to instal Jewish settlements on the West Bank side of the frontier zone. Seventeen such settlements have been established in the past five years. It is planned to set up nine additional settlements in this zone in the near future. This is, in fact, part of a programme to build numerous Jewish settlements in various parts of the West Bank, (see Rowley, Chapter 15, this volume).

These developments have set into motion growing political activities among the Arab inhabitants of the West Bank which have already had some bearing on the interaction between the inhabitants of the West Bank and Israeli sides of the Green Line frontier zones. There are initial indications that a new phase in relations and activities is evolving along the Green Line. These will have to be examined over the next few years before any conclusions on their trends can be reached.

FOOTNOTE

1.   This study is mainly based on fieldwork carried out in 1967-72 and 1977-80 by staff and students in the Department of Geography at Tel-Aviv University under the author's guidance.

REFERENCES

AUTHORITY FOR ECONOMIC PLANNING [1967]
    The West Bank - An Economic Survey, (Government of Israel, Jerusalem).
BAR-LAVIE, Z. [1981]
    The Hashemite Regime 1949-1967 and its Status in the West Bank, (Shiloah Institute, Tel-Aviv), (in Hebrew).
BRAWER, M. [1978]
    'Impacts of boundaries on rural settlement: the case of Samaria', Geo-Journal, 2, 539-47.
BRAWER, M. [1980]
    The 'Green Line' - the Boundary of the West Bank, (Tel-Aviv University, Tel-Aviv), (in Hebrew, English resume).

COHEN, A. [1965]
Arab Border Villages in Israel, (Manchester University Press, Manchester).

COHEN, S.B. [1978]
Jerusalem - Bridging the Four Walls, (Herzl Press, New York).

FRAMKE, W. [1968]
'Die deutsch-danische Grenze in ihrem Einfluss auf die Differenzierung der Kulturlandschaft', Forschungen zur deutschen Landeskunde, 172.

HOUSE, J.W. [1959]
'The Franco-Italian boundary in the Alpes-Maritimes', Transactions, Institute of British Geographers, 26, 107-32.

KARMON, Y. [1977]
'The influence of the Green Line on two neighbouring Arab villages', in D. Grossman and A. Shmueli (eds.), Judaea and Samaria - Studies in Settlement Geography, Volume 2, (Jerusalem), (in Hebrew).

KENDALL, H. [1949]
Village Development in Palestine During the British Mandate, (London)

LIFSCHITZ, J. [1970]
The Economic Development of the Occupied Territories 1967-1969, (Maarakhot Publications, Tel-Aviv), (in Hebrew).

MINGHI, J.V. [1963]
'Boundary studies in political geography', Annals, Association of American Geographers, 53, 407-28.

PRESCOTT, J.R.V. [1966]
The Geography of Frontiers and Boundaries, (Hutchinson University Library, London).

SCHWIND, M. [1950]
Landschaft und Grenze. Geografische Betrachtung zur deutsch-niederlandische Grenze, (Bielefeld).

SCHWIND, M. [1962]
Deutsch-niederlandische Begegnung im Raum des Bourtanger Moors, (Kiel).

SHMUELI, A. and I. SCHNELL [1983]
The Arab Villages of the Little Triangle, (Tel-Aviv University, Tel-Aviv), (in Hebrew).

SOLCH, J. [1923]
'Geographische Krafte im Schicksal Tirols', Mitteilungen der Geographischen Gesellschaft in Wien, 6.

VERHASSELT, Y. [1966]
Les Frontieres du Nord et de l'Est de la Belgique, (Brussels).

WILKINSON, H.R. [1955]
'Yugoslav Kosmet; the evolution of a frontier province and its landscape', Transactions and Papers, Institute of British Geographers, 21, 171-93.

Chapter 14
DUALISM AND LANDSCAPE TRANSFORMATION IN NORTHERN SINAI — SOME OUTCOMES OF THE EGYPT-ISRAEL PEACE TREATY

Nurit Kliot

INTRODUCTION

A place is a piece of the whole environment which has been claimed by feelings. A place is a given space that can be directly experienced, intimately known and passionately loved by its inhabitants (Lewis 1979: 40-41). All places have a sense of shared experiences and often that experience is not shared by others who do not inhabit that particular place. Those qualities which invoke a sense of security, familiarity and of habit of belonging, are sense of place qualities (Wales 1979: 74). A sense of place implies that a community has an accurate reading of its past and present, and some idea regarding its future.

The essence of place is threefold: the first element in the essence of place is the meanings we attribute to places. The identity of places is largely an expression of communally held beliefs, values, and of interpersonal involvements. Second are human activities in a place. Places are moulded by people, and there is a continuous change in places. But people change continuously with places and their experiences in a place tie them deeply to that place. Ties to a place increase as the interaction with others increases. The relations between a community and place are very strong and each of them strengthen the identity of the other. Third, there is the physical base or natural landscape of a place. Most places are located somewhere and have a site but their location is neither a necessary nor a sufficient condition of a place (Relph 1976: 42-44). Places emerge or become through our living in them. Inherent in sense of place is the notion that rootedness and care are inseparable from it. What happens when the process of creation of place is halted by two governments who decide to sign a peace treaty which will result in the transfer of a territory from the one to another? What happened to the sense of place amongst the inhabitants of Yamit, in Sinai, who had a past, a frightening prese nt and no future? Is it possible to lose a sense of place or replace it? Is there any chance that the inhabitants of one place could duplicate that place and their own sense of place in another location? What happens to people with a lost sense of place? These are some of the questions that will be discussed in this paper.

This paper presents the preliminary findings of a long-range study. It is based on long interviews with 60 families from five of the rural communities in the Yamit region. The settlers comprise a cross-section of the Israeli society. They came to the Yamit region from kibbutzim, from metropolitan areas and from other moshavim. Most of them were Israeli born but some of them were immigrants from U.S.S.R., U.S.A. and the Republic of South Africa.

THE CREATION OF A PLACE: VALUES, IDEOLOGIES AND BELIEFS AND THEIR IMPACT ON PLACES.

Landscape as a clue to culture and ideology is a part of culture. Meinig has pointed to the ideology manifested in the American landscape. Important values in the American ideology such as freedom, individualism, competition and progress are expressed in this landscape (Meinig 1979: 33-40). Cultural landscapes are shaped and moulded by 'authors' who are difficult to identify because landscapes are the products of pluralities, rather than particular authors (Samuels 1979: 70-75). Thus, our human cultural landscape is our written autobiography, reflecting our tastes, our values, our aspirations and even our fears. It is also important to note that many authors of landscape are elites and choice is the central criterion of elitism.

The settlement plans of Yamit region and the performance of these plans was a public-government project. Settlement policies and settlement projects have always been a public domain in Israel and basically motivated by ideological principles. The first ideological principle (perhaps a myth) can be called the 'security myth'. Basically it states that Jewish settlements, wherever located, will secure the future boundaries of Israel. Both strategy and tactics of the settlement policies were founded on that principle. As it turned out, this policy was successful in changing the assigned territories for the future Jewish state in the various British plans for the division of Palestine (Oren 1979: 24-27). Land acquisition policies were also formulated along security needs and preferences (Reichman 1979: 49-59). Relatively speaking, this policy was successful: only nine settlements were conquered by the Arabs and had to be evacuated in the War of Independence of 1948. The events of Tel Hai in the 'Galilee Finger' and the security settlements of the 'Tower and Stockade' type, together with the heroic struggle of settlements in the War of Independence, provided the final stamp to the mythical conviction that 'where Jewish settlements are, where a Jewish plough ploughs the soil there will be the boundaries of the state' (Rogel 1979: Zur 1980; Ben-Gurion 1976: 517-526; Allon

1981: 81-83). A secondary myth, which is connected to the
security myth can be termed the 'Sacredness of Settlement' and
it states simply that settlements cannot and will not be removed
from any area of the land Israel ('Eretz Yisrael') because
settling the land is a sacred mission (S. Peres, in Introduction
to Schiff (1979).

A second cornerstone in the Israeli ideology is the 'conquest
of the desert' and its accompanying image of ascetic pioneers
labouring in the harsh environment (Waterman 1979: 176-77). The
ideal of conquest of the wilderness and the pioneer labouring
the land by manual work was a major component in the socialist
pioneering ideology which constituted, more or less, the national
Israeli ideology during the 1950s and early 1960s (Eisenstadt
1967: 13-15). The epitome of this ideology was the mythical
call to 'Make the desert bloom'. The Israeli perception of a
desert land should be paralleled to the human reaction to ruins.
Jackson points to the role of ruins as an incentive for restoration
and for a return to the origins. The joy and excitement involved
are not so much out of creating the new, as of redeeming what
has been neglected, and this excitement is particularly strong
when the original condition is seen as holy or beautiful (Jackson
1980: 102). The desert in Israeli terminology is not identical
with a geographical definition of desert and should be read as
wilderness. Wilderness is broadly defined as any desolated
wastelands, mostly uninhabited by Jews. Thus, the swamps, the
arid land of the Negev and the mountains of the Galilee are
'desert' awaiting conquest of the lands of the pioneer.

The Southern Project Plan which was responsible for the
settlement of the Yamit region was based on the same ideological
principles. Its first goal was to create a populated security
buffer zone between the Arab Gaza Strip and the Sinai Peninsula.
The buffer zone was to be connected to the settlement area in
the Negev and was to form a dense belt of settlements (Nachmias
1977: 9-13). A second aim of the plan was to base the new
settlement area on export-oriented winter farming, and a third
goal was to create an Israeli pioneer capable of making the
desert bloom and of earning his living from his own labour
(Nachmias 1977: 10).

The first call to settle the Rafiah Approaches appeared in
the Israeli newspapers in 1970 and was spelled out in a well-
known ideology-loaded terminology. The newspaper advertisement
and the brochure which was sent to the applicants, emphasised
the challenge of a new settlement project in the new territories
and elaborated on the 'deep experience of establishing a new
settlement in an area not yet touched by a Jewish hand' (Shaflan
1981: 14).

According to their own testimony, the settlers of the Yamit
region were motivated by the pioneering challenge and by the
fact that they were fulfilling a security mission for the state
of Israel. In all the settlements with no exception, the first
year or two were very hard and the physical environment very

hostile. The settlements were remote, and their communication with the rest of Israel - bad. The settlers felt that they were creating something new and their self-image was confirmed by the visits of political leaders such as Golda Meir and Moshe Dayan who called the settlers 'Pioneers' and the media who glorified their hardship. Most of the settlers were fully aware that they were settling beyond the Green Line, but were not concerned at all - the history of Zionist settlement in Palestine showed that boundaries were determined by settlements. They never had any doubts, second thoughts or speculations on their future in the Rafiah Approaches, and perhaps this is the reason for the total shock and confusion when the government and the Knesset decided on a complete settlement evacuation from this area. A process of dis-idealisation began and settlers became confused. They had to find a way to reason and rationalise why they were pioneers no more and why there was no need for a security buffer zone. But the most difficult reconciliation to make was between 'peace' as a supreme value of Israeli society and the need to desert home, fields and places which they created with much love, caring and effort. Many settlers were not able to reconcile the two even by the time of evacuation (April, 1982). Many became hostile towards the government, the Knesset and the settlement organisations. Thus they defined the evacuation as a 'holocaust' and themselves as 'refugees', 'uprooted and 'evacuees'.

They are certain that no settlement in Israel can be sure that one day it will not be uprooted, also. All their values and beliefs have collapsed and their distrust of the Israeli Government is large. Almost all the settlers are convinced that no real peace will evolve between Israel and Egypt because 'Israel has given all and Egypt has given nothing'. None of the settlers was ready to re-settle beyond the Green Line and some are looking for farmsteads in established moshavim where they will not have to start from the beginning. The settlers of Northern Sinai already had an emotional experience in returning a place when the vegetable garden of Neot Sinai was returned to Egypt, together with the El-Arish region in 1980, after overt conflict between the settlers and Israeli soldiers. It is proper to end the discussion on the ideological facets of place by a quotation from Ezer Weizman (then Israel's Minister of Defence), spoken in the Israeli court where the same settlers stood trial:

'The settlers accused are exceptional youth who came to the Rafiah Approaches after the Six Day War, worked hard and turned their settlement into a garden in the desert. Their refusal to turn their property over to the Egyptians is ideologically motivated' (Weizman, testimony given in the military court of Gaza, 12.5.80).

ACTIVITIES AND BEHAVIOUR SETTINGS AS MAKING A PLACE

Places are made by human beings and their activities. The concept 'behaviour setting' can express well the connection between a place, activities and a group of people (Perin 1970). The most important behaviour setting in the Yamit region is being an agriculturalist. Farming is perceived as a way of life and not only as a way to earn a living. The farming areas and the farmsteads manifest a designed landscape because every community was planned to cultivate certain crops and particular technologies to achieve the farming goals. Thus, the planners of the Jewish Agency had an impact on farming and on a sense of place, a result of certain farming methods. The communities which were designed to cultivate open fields developed a deeper sense of place and greater attachment to the land. Partially, the reason for this was the exceptional agricultural achievements of those communities, achievements which soon were translated into economic prosperity. The communities which were planned later, were based on glasshouses which were found to be less successful and were nicknamed 'the Glass Ghettoes' by the less successful farmers. The two oldest communities, Sadot and Netiv-Ha'asara, which were based on open fields became symbols of success and as an example towards which to aspire - and perhaps - to envy. But the nature of open field farming dictated an extensive use of cheap labour, especially in harvesting the vegetables and flowers. By their own social principles moshavim such as Sadot and Netiv-Ha'asara had to rely only on their own labour and the broad use of cheap labour, mostly Beduins who were evacuated from the area, brought the moshavim into a sharp conflict with the settlement organisations and the government. The glasshouse farming was developed as an answer to the problem of extensive hired labour, but as noted earlier, their economic achievements were poor. Thus, farming has contradicting features: the pride on achievement, the economic success, a rapid process of rootedness but also a sacrifice of an important social principle. The image of the 'pioneers from Rafiah' rapidly deteriorated in Israeli society (as expressed in the media) and as a result, the self-image of the settlers themselves changed with some of them attempting to reduce the use of Beduin labour as much as possible. Farmers turned to new crops and branches in which outside labour is not needed. The sandy soils of the Yamit region were found to be most suitable for modern farming technologies such as drip-irrigation. The settlers of Yamit region were anxious to experiment and try new methods and technologies in order to succeed and their success, so they feel, is irreversible in any other place.

A second behaviour setting derives from the social principles of the moshav which advocate mutual assistance and mutual guarantees among members of the moshav. In all the communities this principle was applied extensively. Members assisted in farming when a certain farmer was sick or in the military reserves and sick housewives were assisted by friends in all household work. Another custom which enhanced communal ties was the adoption within the moshav of new families by veteran families. Adoption meant that the newcomers were welcomed and that the community was doing its best to absorb the newcomers in the easiest way possible.

While mutual assistance and mutual responsibility were reflected in the social life of the Moshavim, the cultural events strengthened the communal ties. Holidays such as Passover, or the 'Moshav Holiday' which symbolised the community anniversary were celebrated within each Moshav, integrating all the members of the Moshav. But some cultural events were specially designed for all the Yamit region settlements. The regional celebrations of Independence Day or the regional Bar Mitzva celebrations became a reflection of the communal spirit of the region. The fear of losing the 'togetherness' within communities caused many settlers to seek a common solution to their re-settlement. Many of them planned on creating new communities in the Israeli Negev in order to preserve the Yamit social and cultural patterns. Many of the settlers stated clearly that they could not separate from their friends in the Moshav and with them they were ready to go to the 'other end of the country'.

SENSE OF PLACE: SYMBOLISING PLACES AND PROVIDING THEM WITH MEANINGS

Symbols are living and concretised ideas of meaning which unite the members of a group on the various levels of their existence (Solomon 1955: 103). Symbols are part of the process whereby the experienced world, the world of perception and concept, is created out of the physical reality (Lee 1954: 73-75). The most important symbols are monuments of different kinds. Monuments are numerous in any landscape where inhabitants share a strong sense of religious or political past and moreover, are concerned with their beginnings (Jackson 1979: 91-92). The monuments of the Yamit region reflect the fact that the region has had a short history, from the Israeli settlers' point of view. The military history of the region gave birth to all of its monuments: in the town of Yamit a monument to the 180 dead soldiers of the 'Steel Division' is located in the centre-to-be of the town that shall be no more. In nearby Sadot, a monument comprised of trees and cement structures commemorates the dead heroes of a particular troop that fought there. A third monument was a reminder of an accident involving an Air Force helicopter in which ten soldiers were killed. A nearby public beach was also named after those ten soldiers. The specific military history was also expressed in a more abstract way: in the names of two settlements in the region: Ugda which is the Hebrew name for an army division: and Netiv-Ha'asara ('The Path of the Ten') which also commemorates the ten deceased soldiers.

178

The Israelis are not obsessed by the military, but the reality of participating in four major wars since independence, creating a situation in which almost each community has lost sons in one or more of the wars and in each community a building, a garden, a library or a playground commemorates these sons. Those monuments are called 'Yad Lebanim' ('Memorial for the Sons') and they mirror the more general nature of remembrance towards those who fought for the state. The ceremonies of 'Memorial Day' take place in the 'Yad Lebanim' and the continuity between the living and the dead is thus ritualised. The settlements of the Yamit region are young and none has lost a son in any of the wars; yet, a 'Yad-Lebanim' monument was constructed in the centre of Sadot, nearby an old sycamore, and it served all the settlements as the gathering place for commemoration on Memorial Day. In that respect one can see the erection of a monument as a necessary layer in the process of place creation. The Yad-Lebanim of Sadot symbolises the ties to the general Israeli society and to its symbols and is an essential part of the Israeli identity and an additional cord to tie a man with his land.

In the process of the symbolic interaction between man and land, rural and agricultural symbols become very important. It began with the selection of place names having rural connotations and it continues with a process of attributing symbolic meaning to each planted tree and grove. When members of Sadot chose the name 'Fields' for their community there was no sign in the surrounding desert scenery that the name would ever be suited to the landscape. But in the short span of a couple of years, green fields surrounded the settlement and in 'the gates of the desert a paradise was blooming' in the excited words of one of the commanders who fought in this region. Agricultural meaning is found in the names of 'Priel' ('God's Fruit') and 'Prigan' ('Garden's Fruit'), or with the prefix 'Niv' ('field') another name. The natural landscape impressed the settlers who referred to the beauty of the palm-groves by names such as 'Dikla' ('Palm') or 'Neot Sinai' ('The Oasis of Sinai') and by names such as 'Holit' ('Dune') or 'Sufa' ('Storm'), which could be interpreted as very fast and could probably hint to the military conquest of Northern Sinai.

The creation of greenery around the settlement was very difficult because of the free movement of the dunes. 'Conquering the desert' and 'making the desert bloom' were no more just plain metaphors, but a real struggle. This explains at least partially, the pride, reverence and emotions involved in every tree, lawn and rose bush which was rooted and which flourished. The trees, lawns and flower beds did stop the moving dunes and the sand storms and also became a source of aesthetic pleasure and pride. On the light side, 'competitions' were set between newborn children and newly planted trees to see whose would grow faster; yet the deep involvement of the settlers with their man-made environment resulted in extreme emotions when their uprooting was decided. One reaction was to kill all greenery and let the

179

desert rule again, and another was to dig up the trees and transfer them to the new home or place of residence wherever that was to be. The rationalisation of these two modes of behaviour was one: 'I have planted the trees, I fought the cruel desert until my trees won, I have created a garden and a place, and no one can take that from me.' Thus, place will become a non-place (a desert) again or a place will be re-created by moving the personal trees to the new home area. The image of Egyptians enjoying one's own garden and trees is blasphemous and profane. It is an image in which the act of strangers enjoying one's own garden was considered a hostile and aggressive act.

We know that every new revolutionary social order is anxious to establish its image and acquire public support and in order to achieve these goals it produces many commemorative monuments and symbols and encourages public ceremonies around them. The Yamit region was inhabited mostly by a nomadic population with hardly any visual impact on the landscape in that respect. The process of creation of place by the Israelis added monuments to the landscape and one of the fears expressed by settlers and bereaved families whose members are commemorated in those monuments is that the Egyptians will not respect the monuments; hence, the request of some of them to uproot them from their original sites and replant them in Israeli territories.

It was decided that Sadot will move the 'Yad Lebanim' monument and the monument nearby to the site where the largest group of moshav members decided to resettle. The Yamit monument will be re-located in the Besor region, a nearby settlement region which was intended for the evacuated settlements of the Yamit region. It will be the first time in Israeli history that important symbols are moved to a 'safe' territory.

THE NATURAL LANDSCAPE AND SENSE OF PLACE

The geographical inventory of the Yamit region can be presented as follows: arid plain covered mostly by sand dunes, and sandy soils. The climate is moderated by the sea and is comfortable for the population. The only climatic flaws are the sand storms in the fall and the very low precipitation - no more than 250 mm of rain annually. The natural vegetation wherever it has not been removed by Beduin livestock, is desert scrub and bushes. Two acquifers along the coast in which the water level is high were turned by the Beduins into date groves and to small fruit orchards and vegetable patches.

The attachment of the inhabitants to the natural environment is complex and has a dualistic nature. Three landscape elements were identified by the inhabitants as major components of the visual environment: the sand dunes, the sea and the palm trees. This combination of elements suggested the uniqueness of the landscape. Another characteristic of the environment was that it suggested participation. Participation means that the

180

environment offered engagement, both physical and mental. Thus adults and children rolling and playing in the sand dunes and walking with bare feet on the sand dunes was a frequent experience. The sea and the beautiful harbours also invited a very active participation. The last characteristic of the natural landscape is its scale which is regional for all the settlers. All were well acquainted with the whole region and knew well any beautiful corner in that landscape.

The sand and sand dunes were described as white, yellow, gold, smooth, hypnotising, but also as dull and monotonous. The sand dunes became, after a short period of experience, a threat – the sand storms became a nightmare to any family because there was no hideaway. The dunes presented a continuous challenge for the farmer: how to turn these yellow carpets into green carpets. It seems that the promise inherent in the yellow dunes was most intriguing for the inhabitants and once they know from their experience the agricultural success embedded in the golden dunes, the meaning of dunes became complex and cultural in its content, more than purely natural as it was before. The sand dunes became a symbol of the continuous struggle of man to conquer the desert and make it bloom and the primeval landscape a challenge for the pioneering spirit. The sea and the beaches were the second landscape feature which impressed the settlers and the sea is very pleasing view even for those who do not go often to the beach. Their pleasure is sensual and visual more than participatory. The palm groves which create a long oasis along the coast were described as 'picturesque', 'exotic' and as 'tourist havens'. The image of the combination of the blue sea, yellow dunes and green palm trees was strong and vivid and all the major vistas people had where composed from those components. The most important vista is that of the entrance to Yamit when suddenly the sea and the palm trees are uncovered to the traveller.

Though the Yamit region was new for most of the settlers they saw it as Israeli Landscape. Some settlers said that it is part of the Negev, the southern (desert) part of Israel, while others compared it to their home environment and saw similarities in the landscape. Thus, people could identify more with Yamit because they imagined it as a similar landscape to their home community such as Mikhmoret or Sdot-Yam which are coastal communities. People also tied Israeli Landscape to the region by symbolic actions such as bringing stones, pieces of wood, and soil from their home community to their new home in Yamit. The comparison with the Israeli landscape continues in other aspects: people noticed that there is no mud in the region and mud was inseparable from the life of a farmer in all other parts of Israel. It took a while to get used to the reversed agricultural seasons. In Northern Israel, the spring and summer are the busiest agricultural seasons while in the Yamit region the fall and winter are busiest, while in the spring and summer most of the area looks yellow and dry. Few people mentioned that they missed the view of the green hills in Northern Israel and more particularly - the view of trees and forests. Greenery was a missing element in Yamit and people never had enough greenery. Perhaps this is the key for the understanding of the meaning of the natural landscape to the people who settled it. This landscape was mostly characterised as a desolate land or as a wasteland and wilderness waiting for human occupance. The image is of man as an active agent in that wilderness and the natural landscape becomes really attractive (in the eyes of the settlers) when it is green, when trucks load fresh tomatoes in the middle of the winter and when the settlements form green islands in the yellow spaces. From all the components of the physical landscape the residents will miss mostly the sea - not necessarily as a participation site but as an aesthetic view. The pleasantness of the weather is the second element that will be missed. But the residents of Yamit left no doubt that their attachment to their environment was firstly and most importantly its cultural contents.

LOSING A PLACE AND THE FEAR OF LOST SENSE OF PLACE

> Lela 'Your shock is caused by the fact that your home is totally ruined. One moment you are with your roots deeply in the land and in another you are uprooted'.

Little is known or written on the absence or lost sense of place. Lewis states that the lack of a sense of place means a lack of responsibility to place and the consequences are horrendous (Lewis 1979: 34-35). Others are convinced that people can get attached very rapidly to new places if these are similar to those already known to them. Relph quotes Ian Nairn who claims that people root themselves rapidly and mobility strengthens the sense of place (Relph 1981: 14).

The decision to evacuate the settlements in the Yamit region caused a total collapse of the world of the settlers. Suddenly their home territory no more provided security and opportunity. The forced evacuation started a process of community disintegration - more spiritual than physical. While few people left the region physically, social and cultural life stopped and people did not find the necessary energies to continue with the regular patterns of culture, the togetherness which typified the cultural and social activities of the communities disappeared because people became preoccupied with a new behaviour setting: a search behaviour for a future.

It is suggested that the process of place-creation should be perceived as a cycle. Place-creation cycle is measured by generations, not by years and the place creation cycle of the settlers of Yamit was halted by the Peace Treaty Israel signed with Egypt. Cutting off the process of creating a place produced different attitudes towards one's own place: first, an attitude

Figure 14.1: The Settlements of Northern Sinai and the Besor Region 1973–82

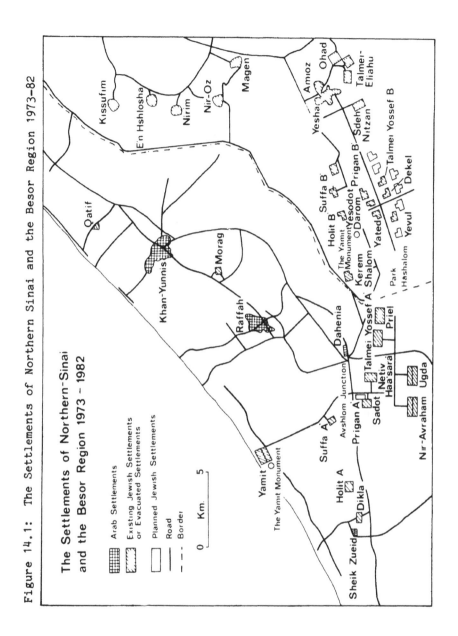

of negligence and uncaring for the home and its environs. A small group of settlers, mostly non-achievers and sometimes losers, who failed in farming, did not develop a deep attachment to their place and the Peace Treaty with Egypt provided them with an easy exit from their communities. Second is the attitude of a small group of settlers, members of the 'Movement Against the Retreat from Sinai'. The members of this group did not sign any reparations agreement with the government until very late and believed that Sinai should not be returned to Egypt. This group was continuing the process of place creation: they continued farming and planted trees as a manifestation of their strong convictions that they will be in Sinai to enjoy the fruit of these trees. For them, it was necessary to complete their place creation in Yamit, and they were deeply convinced that the Israeli society as a whole would pay dearly for the evacuation of settlement. This group was highly alienated from the central government and they felt themselves betrayed by it.

The majority of settlers comprised a third group which revealed, in its behaviour, patterns which typify the above groups. It is interesting to note that the threat of losing 'their place' already produced emotions of nostalgia and people found themselves walking to different spots and examining their feeling of attachment to those spots. They had already begun the process of 'idealisation' of a place, a process which generally appears a long time after a place is deserted. Accompanying the feelings of nostalgia were feelings of outrage that their homes and gardens would be left for the Egyptians. Most of the settlers stated that they would destroy their own house and gardens and would sterilise the soil, so that would never yield again. No one is indifferent to the image of 'strangers using their homes'.

If a place creation is a cycle, one can expect that the settlers of Yamit will seek a way to repeat the process or re-create a new place for their own. The search for a new place, where one has to re-plant himself, is typified by two principles, replication and duplication. Replication means the need to recreate a place with one's own friends. The loyalty to the other evacuees almost resembles a tribal loyalty and people feel that only with 'their own friends', those with whom they grew together, could they (perhaps) create a new place. This wish of the people to continue with their own community has resulted in the creation of two new settlements in the Israeli Negev − near Magen and near Erez. These settlements are open only to the Sinai evacuees and the settlers are constructing those settlements with the compensation money received for their homes and property. The other characteristic of the search for place duplication means a need to find a similar place. The settlers who looked for farmsteads on established moshavim in Northern Israel, looked for an environment which they perceived as similar to their own home. Perhaps this is the reason why so many of them looked towards re-settlement in the Negev which is perceived

as similar to Yamit, or for a farmstead in the Western Galilee
with its combination of sea and dunes. Even the new sites for
the settlement which are to be created by the evacuees themselves
were selected largely on their perceived sameness to the former
home surroundings. Another aspect of duplication of a place is
expressed by a wish to transfer precious things like trees to
their new homes. To weave a cord of continuity a person will
move a piece of landscape, his communal behaviour settings, his
memories and his symbols (such as monuments). These components
may assist him in finding his lost sense of place. Removing
trees and monuments manifested more than any other action that
the metaphor of growing in a place, taking roots in a place,
rootedness - is not metaphor at all but inherent in the creation
of place. Man and plant both find difficulties in being replanted
and rerooted. There is exhaustion in the need to 'start all
over again' because so much energy and self was wasted in the
process of place creation. A process of duplication is the essence
of losing a place. One needs to complete his place creation
cycle either in his own place or in a newly-created place and
to do so he needs to recreate - socially, culturally, and
environmentally - the same place.

REFERENCES

ALLON, Y. [1981]
Focal Points, (Hakibbutz HaMeuchad, Tel-Aviv), (in
Hebrew).
BARKER, J. [1980]
'Designing for a Sense of Place in Mississippi', in P.
Prenshaw and J. McKee (eds.), Sense of Place, (University
of Mississippi Press, Jackson), 162-78.
BEN-GURION, D. [1971]
Memories, (Tel-Aviv), (in Hebrew).
EISENSTADT, S.N. [1967]
Israeli Society, (Weidenfeld and Nicolson, London).
JACKSON, J.P. (ed.) [1980]
The Necessity of Ruins, (University of Massachusetts Press,
Amherst).
LEE, D. [1954]
'Symbolization and value', in L. Bryson et al., (eds.),
Symbols and Values, (Harper and Row, New York), 73-86.
LEWIS, P.F. [1979]
'Axioms for reading the landscape', in D.W. Meinig (ed.),
The Interpretation of Ordinary Landscapes, (Oxford University
Press, New York), 11-32.
LEWIS, P.F. [1980]
'Defining a sense of place', in P. Prenshaw and J. McKee
(eds.), Sense of Place, 24-46.

MEINIG, D.W. [1979]
    'The beholding eye', in D.W. Meinig (ed.), Interpretation,
    33-50.
NACHMIAS, D. [1977]
    The Southern Project: Final Report, (Settlement Study
    Centre, Rehovot) (in Hebrew).
OREN, E. [1978]
    The Settlement in the Years of Struggle, (Yad Ben-Zvi,
    Jerusalem), (in Hebrew).
REICHMAN, S. [1979]
    From a Territorial Holding to a Settled Land, (Yad Ben-
    Zvi, Jerusalem), (in Hebrew).
RELPH, E.C. [1976]
    Place and Placelessness, (Pion, London).
RELPH, E.C. [1981]
    Rational Landscapes and Humanistic Geography, (Croom Helm,
    London).
ROGEL, N. [1979]
    Tel Hai, (Hadan, Tel-Aviv), (in Hebrew).
SAMUELS, M. [1979]
    'The biography of landscape', in D.W. Meinig (ed.),
    Interpretation,
SCHIFF, Z. [1979]
    War Without End, (Schocken, Tel-Aviv), (in Hebrew).
SHAFLAN, Y. [1980]
    Settlement in the Rafiah Region, (in Hebrew).
WALES, R.W. [1980]
    'Environment, a sense of place and public policy', in P.
    Prenshaw and J. McKee (eds.), Sense of Place, 162-78.
WATERMAN, S. [1979]
    'Ideology and events in Israeli human landscapes',
    Geography, 64, 171-81.
ZUR, Z. [1980]
    Settlements and the Boundaries of the State, (HaKibbutz
    HaMeuchad, Tel-Aviv), (in Hebrew).

186

Chapter 15
SPACE, TERRITORY AND COMPETITION
— ISRAEL AND THE WEST BANK

Gwyn Rowley

Whereas the first five books of the Bible, the Torah (Pentateuch), present an essentially theocentric viewpoint, for Jews they nevertheless establish and pledge allegiance not to land or a land but to the land (Rowley 1981). Thus in the Torah there is no timeless space nor is there spaceless time. Rather there is storied place, that is a place which has meaning because of the history lodged therein. In essence, the unbreakable dialectic between a people and a place increases the burden of awareness (Hall 1966; Tuan 1977), Smith suggesting that: 'It is briefly history that makes a land mine' (Smith 1969: 109), while Brueggemann (1977: 5-6) asserts that the land assures Israel of its historicality and is a repository for identity.

Jacobson (1982) likewise focuses upon this question of land and demonstrates its centrality to his recent essay, which concerns not only the Chosen People and the God who chose them but also, and necessarily, considers the notion of the Promised Land and the God who promised it. For Jacobson the question of land - of losing land, regaining land, losing it and repossessing it - is central to the story (see Rowley 1981: 462). Although there are problems concerning the factual nature the precise historicism, and interpretation of these historical events, (Eichrodt 1961 Vol. 1: 37; Newman 1962: 29-38; Rowley 1981: 444-48), it should be realised that here we are dealing with a highly subjective view of history (Daiches 1975: 7)

National memory! Although nationalism is considered usually as a quite recent, post-Renaissance development, the Hebrew nation arose from a clearly defined consciousness of being different from other peoples, the Gentiles. This national character and enduring spiritual creative energy of the people came to form a cultural continuity that was to prove stronger than racial, political or geographic continuity, and Israel in those far- off times came to represent the strong consciousness of a cultural mission, possessing three of the essential traits of modern nationalism: the idea of the Chosen People (covenant), the emphasis on a common stock of memory of the past and of hopes for the future (destiny) and, lastly, national messianism.

At a philosophical level this study can be set into a broader field of ethological inquiry. For example, as with most social animals, man has been shown to seek security at the centre of his territory when he is frightened and vulnerable (Hediger 1949; Fischer 1971; Leyhausen 1971; Esser 1971; Tuan 1979); he will often seek dominance within it and in relation to the territory of others (Wynne-Edwards 1962; Calhoun 1971; Lorenz 1967; Leyhausen 1971); and will exhibit exploratory or aggressive behaviour when motivated by need or excess of free energy (Hediger 1949; Leyhausen 1971; Calhoun 1971; Ardrey 1966). He may do so in terms of physical and/or conceptual territory. Yet, as considered by Tuan (1979: 6), metaphysical terror, unique to human beings, cannot be assuaged anywhere in this world. Only God may provide relief: 'Rock of Ages, cleft for me; let me hide myself in thee' is indeed a desperate plea. Tuan identifies 'landscapes of fear' as a construct of mind as well as physical and measurable entity, relating both to psychological states and to tangible environments. Likewise for the Israelites 'wilderness' signified a demonic, ungodly phase.

This veritable desert, a waste land, conveys the picture of spiritual emptiness! In addition to what may be referred to here as the theological bases of Zionism we will elsewhere consider the Reconstructuralist approach that viewed Zionism as a bastion against Jewish assimilation in the supposedly liberal enlightened post-Darwinian Europe of the 19th century. In essence Reconstructuralism viewed Jewish 'peoplehood or nationhood' rather than religion as the central aspect of Judaism (Rowley, forthcoming).

ERETZ YISRAEL

Eretz Yisrael, the Land of Israel, the Promised Land, is essentially the antithesis of and a derivative construction from a landscape of fear with its related inchoate experiences and doubts. As we have demonstrated elsewhere the covenant between God and the Israelites must be viewed as a deed of title with solemn obligations on behalf of both parties; in essence land for obedience (Rowley 1981). This conditional nature of the covenant has to be stressed. Possession of the land is and must always be conditional. Thus for a landless, fearful people wandering in 'The great and terrible wilderness' (Deuteronomy 1:19) Eretz Yisrael came to represent a vision of a landscape of security and God-given protection, permanent settlement and comfort, hope and promise.

In the context of this essay, however, we will see that the notion of Eretz Yisrael has always included Judaea and Samaria and especially Jerusalem within the general territorial schema. This is demonstrated in Isaac's (1976: 27-29) review of the development of what she terms 'Normative Zionism' and Rowley's (1981: 444-46) brief consideration of the background to and the emergence of the 'covenantal promise' with the notion of Eretz Yisrael; in the words of Theodor Herzl 'the Promised Land for a Chosen People'. The actual territorial extent of Eretz Yisrael is considered in a number of verses within the Old Testament. Scripture is generally less precise concerning the specific boundaries of this homeland, or rather several interpretations can be placed upon the various biblical references to this Promised Land, (Genesis 15: 18; Numbers 34: 3-12). A fundamental point is that the notion of the Promised Land has always included Samaria and Judaea, the presently so-called West Bank including Jerusalem, the areas being firmly embedded in the twelfth- century B.C. lands of the Twelve Tribes, with Israel reaching its maximum extent during the reign of King Solomon in the tenth century B.C. The generally accepted territorial extent of Eretz Yisrael now appears to be the Jordan valley to the east, the Israeli-Lebanese border to the north, and a line from Wadi el-Arish in the south-west across to a point to the south of Eilat on the Gulf of Aqaba in the south.

The particular importance of Jerusalem to Israel, both ancient and modern, cannot be over-emphasised. To the religious Zionist whereas land is covenantal this is nowhere more so than within Jerusalem itself. Indeed since the time of David it has been the Jewish ambition to lay particular claim to this single symbolic place. The words of Ben-Gurion in 1949 still represent the prevailing Israeli sentiment concerning the Holy City and the symbolic attachment to land as place with meaning and soul:

> '[Jerusalem] is an integral part of Israeli history in her faith and in the depths of her soul. Jerusalem is the 'heart of hearts' of Israel ... A Nation which over 2,500 years has always maintained the pledge vowed by the banished people on the rivers of Babylon not to forget Jerusalem - this nation will never sanction its separation. (Moreover), Jewish Jerusalem will never accept foreign rule after thousands of her sons and daughters have freed, for the third time, their historic homeland and delivered Jerusalem from destruction' (Divrei Ha-Knesset 1949: 221).

The specific agenda was clear, for present-day Israel to attain its 'birth-right' it would have to seek to incorporate those areas of Eretz Yisrael outside Israel into the developing state, such a spatial expansion being referred to in Israeli parlance as 'the redemption of the land'. Such are the depths of feelings associated with the notion of Eretz Yisrael.

THE WEST BANK AND ADVANCING ISRAELI SETTLEMENT

Particular attention will be centred upon the progressive and continuing Israeli settlement of the West Bank following on from the Six Day War of 1967, for it is this area of 5800 sq. kms., which in 1967 was the home of 650,000 Palestinian Arabs, that would be central to any Palestinian political entity. Following their occupation of the West Bank in 1967 the Israelis embarked upon a policy of colonial settlement within the area. Such settlement is considered to be in direct violation of the Geneva Convention of 1949 which categorically states that:

> 'An occupying power shall not deport or
> transfer parts of its own civilian population
> into the territory it occupies' (United
> Nations 1976: paras. 1-2).

The counterclaim by Israel suggests that the entire West Bank, including Jerusalem, is traditionally within Eretz Yisrael, and thus an integral part of the modern state.

The initial Israeli settlements in the West Bank, mainly agricultural and service centres, were established along the eastern slopes of the Samarian mountains, the Jordan Rift Valley and about the Jerusalem—Latrun salient (Rowley 1981: 457). Between 1971 and 1975 further settlements were developed particularly along the eastern slopes of the Jordan valley but to the west of the earlier settlements, and also in a concentration to the north and east of Jerusalem, and to the west of Jericho and about Bethlehem. The Jordan Rift settlements have now been linked by the new Allon road, a modern rapid-transit military highway (Figure 15.1). The continuing Israeli colonisation of Samaria is of particular importance. Together with its greater reliability of rainfall, Samaria has traditionally possessed a denser population than Judaea and can be generally characterised as a prosperous agricultural area centred upon Nablus (Karmon 1971: 316-17). Whereas the initial Israeli settlements along and overlooking the Jordan Rift Valley could have been considered as military outposts the direct colonisation of the Samarian and Judaean agricultural heartlands around both Nablus and Hebron indicate the permanence and long-term endeavours of Israeli settlement plans for these areas. In essence, we will outline the proposals to settle some 180,000 Israelis within the West Bank, that is Samaria (Shomron) and Judaea (Yehuda) between 1979-83. Such plans should be viewed as a continuation of the Israeli settlement of the West Bank commenced in 1967.

THE DROBLES PLAN

Upon his appointment as head both of the Jewish Agency's Land Settlement Department and the Rural Settlement Department of the World Zionist Organisation in 1978, Matityahu Drobles began to consider formally the development of a consolidated and

Figure 15.1:   Judaea and Samaria, 1978

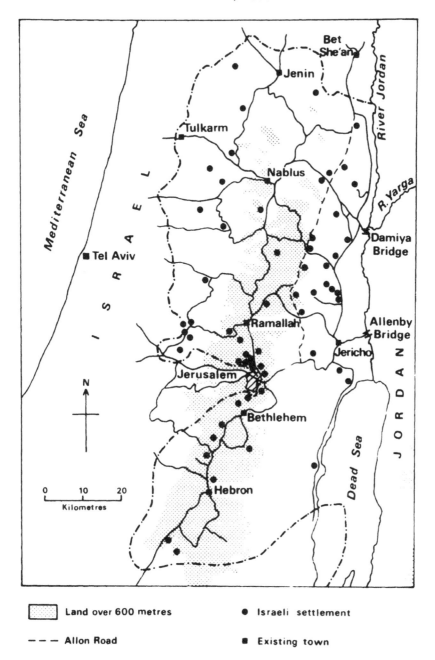

Land over 600 metres   ● Israeli settlement

– – – Allon Road   ■ Existing town

general Master Plan for the Israeli colonisation and settlement of Judaea and Samaria. The resulting Master Plan for the Development of Settlement in Judaea and Samaria, issued in October 1978, (the Drobles Plan), was to cover the five-year period between January 1979 and December 1983 (Drobles 1978).

This Plan seeks to plan for the future disposition and extension of Jewish settlement in the West Bank and its details will be presented and considered here. It has now been accepted as policy by the Likud government, indeed as the specific agenda for colonisation of the West Bank. Fund-raising activities around the world by the World Zionist Organisation are underwriting much of the financial costs of the entire programme (Figure 15.2).

Five pages of text within the twelve-page Drobles Plan provide an itemisation of the specific details relating to the planned 'Disposition of Settlements'. The settlement blocks, together with both the existing settlements where enlargement is planned and the proposed new settlements, are presented in Figure 15.2. Whereas the Drobles Plan is criticised even by some Israelis as being overly simplistic, for example in lacking a finely considered geographic base in the identification and evaluation of the specific sites isolated for settlement, the plan nevertheless represents a prime statement of Israel's resolute intentions to continue to extend Israeli settlement even into the heart of the fertile and prosperous Samarian agricultural region. In fact the specific features of the Drobles Plan link into the overall recommendations of the Begin plan on administrative autonomy for the West Bank and the Gaza Strip presented to the Knesset, (Israeli Parliament), in December 1977. In brief, the Begin plan envisages granting 'administrative autonomy' to the Arabs of the West Bank and Gaza Strip, although the Israeli authorities would continue to be responsible for 'security and public order' and Israeli residents would be permitted to acquire land and settle in the occupied territories.

Settlement Types

The Master Plan itemises the specific nature of the various settlement types. Within urban settlements some 60 per cent of the families will be employed in industry, handicrafts, and tourism and the remainder in services and work outside the settlement. In the towns in proximity to Jerusalem the proportion of those employed in outside work will be greater. In the community settlements the economic basis in the development stage will see about 50 per cent of the families earning their living from industry and handicrafts, about 12 per cent from capital-based intensive agriculture, about 25 per cent from outside work and some 13 per cent from local services. The agricultural and the combined settlements will be based on agricultural branches,

Figure 15.2: Progression of Israeli Settlements in
The West Bank 1967-78

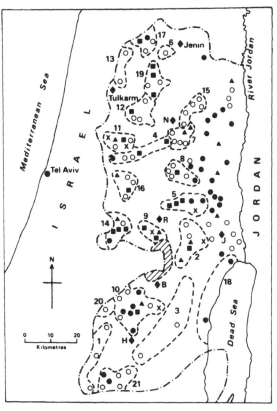

● Settlement founded before
  Begin's Likud government (1977)

■ Settlement founded during
  Begin era but prior to
  Camp David Summit (Sept 1978)

▲ Settlement founded or legalized
  during Begin era but after
  Camp David Summit

O Planned settlement

⟨⟩ Border of settlement block

▨ East Jerusalem
  settlement block

◆ Existing town

X New town

Legend To Settlement Blocks

1	Adorayim	8	Ephraim	15	Tirzah
2	Adumim	9	Givon	16	Neveh-Zuf
3	Amos	10	Gush-Etzion	17	Reihan
4	Ariel	11	Karnei-Shomron	18	Shiloh
5	Beit-El	12	Kdumim	19	Shomron
6	Dotan	13	Maarav	20	Tarkumya
7	Elon-Moreh	14	Modiin	21	Yatir

mainly intensive, depending on the means for production in the area, as well as on industry, handicrafts and tourism. Some of the settlers will be engaged in local and regional services.

The particular features of the community settlement (Yishuv Kehillati) used in the colonisation of the West Bank, require some elaboration. These community settlements are small private enterprise rural settlements. Although the settlers within such settlements have certain communal obligations, such as participation in the settlement co-operative union which has responsibilities for community services and, in an advisory capacity, for economic matters, the settlers are not constrained by the collective decision-making strategies of the settlement and can work in other centres if they so wish. In fact a deal of commuting occurs from Yishuvim Kehillatim to the larger urban places in Israel such as Jerusalem and Tel Aviv-Yafo.

The Drobles Plan emphasises that the settlement:

> 'throughout the entire land of Israel is for security and by right and that settlements within the West Bank have been determined following a thorough examination of the various sites with respect to their being suitable and amenable to settlement, taking into account topographical conditions, land-preparation possibilities etc' (Drobles 1978: 2).

Financial costs of plan implementation

The plan proposes a grand total of 26,855 families for settlement in Judaea and Samaria by 1983, some 9,700 from the enlargement of existing settlements and those that were under construction in 1978 and over 17,000 families within new settlements proposed within the Drobles Master Plan. Multiplying this grand total of 26,855 by the average family investment cost, IL2 million (1978 prices), gives the total of IL54 billion. To this total envisaged cost must be added an additional significant amount to cover inflation and devaluations since 1978 and the cost of further settlements that will doubtless be recommended after ongoing surveys are completed. Allowing for inflation and converting this sum, we obtain an approximate sum of $470 million for the five-year West Bank settlement plan. Thus the minimum, in rounded figures,is $500 million but our conservative yet realistic estimate would now suggest totals of at least $800 million. The World Zionist Organisation is underwriting and maintaining the financial solvency of the entire plan. Whereas estimates may differ from those suggested here the important conclusion is that such a considerable investment indicates not only a real commitment and resolve by the Israelis to settle firmly the West Bank but also, it suggested, an intention to incorporate Judaea and Samaria, when the time is right, into the State of Israel. Indeed, relating to the investment required to finance the continuing settlement of the West Bank we note:

> 'This investment is absolutely essential and is a condition for the execution of a paramount national mission' (Drobles 1978: 3).

New Facts

The intention for permanent settlement is all too clear as Drobles is inferring that the incorporation of Judaea and Samaria into Israel is a realisation of a national mission, that is the unity and integrity of Eretz Yisrael, while the Israeli settlement policy of the West Bank is serving to 'create political facts' regarding Israel's future boundaries in any final agreement with the Arabs (Nisan 1978: 87). While the Israeli policy is to implant agricultural and industrial settlement between existing Palestinian villages the policy includes mitzpim settlements,'observation posts' that:

> 'are intended to be the nuclei of future settlements and are sited in quasi- military fashion in the mountains above the largest Arab villages ...' (Jerusalem Post International Edition, 18-25 November 1979: 13).

The Drobles Plan also includes details of seven planned New Towns in the West Bank. The creation of such towns, together with the settlements considered above, would further reinforce Israeli control and possession of the West Bank by its integration into a remodelled urban network oriented towards and fused into Israel. Furthermore the implantation of these new towns into the West Bank will proceed, albeit slowly, to incorporate the Arab fellahin into the new urban system.

Relating to an autonomy plan for the West Bank, for example as envisaged within the Reagan proposals of September 1982, and the Israeli settlement of the West Bank, the following extract from a Drobles progress report published in 1980 provides further insights into the strategy for the continuing and proposed settlement of Judaea and Samaria to 1985. While Drobles expresses his general satisfaction with progress to 1980, he considers Israel in a race against time. The facts established in Judaea and Samaria between 1980 and 1985 will determine events. Thus he advocates seizure of state-owned and uncultivated lands in order to settle between and around Arab centres to reduce the danger of a [Palestinian] state. He foresees a Jewish population of 120,000 - 150,000 in the West Bank (excluding Jerusalem) by 1985. Drobles' repeated use of the term 'minority' population in his references to the Palestinian Arab population of Judaea and Samaria, which in 1980 totalled c.800,000 as opposed to an

Israeli population of c.25,000-30,000, must be seen to derive from his particular perspective that considers Judaea and Samaria as integral parts of Eretz Yisrael (Drobles 1980: 3-4).

CONFLICT AND THE ALTERNATIVES

On 22 October 1979, the Supreme Court of Israel ruled for the first time that an Israeli settlement in the occupied West Bank, Elon Moreh, established by the Government a few months earlier, was illegal. In doing so the Court judged, in the face of conflicting military opinion, that there was no justification as regards security for the settlement which, unlike earlier settlements, was established in the northern part of the West Bank, densely populated by Arabs. For many right-wing Israelis and particularly for the Gush Emunim religious group which settled Elon Moreh, the Supreme Court judgement is an intolerable precedent, as it suggests that Israeli settlements within the West Bank may be dismantled and their lands returned to their former Arab occupants. Yet, as we have noted, the present Israeli Government is firmly committed to the West Bank.

For many the problems of West Bank colonisation are viewed in essentially dichotomous terms; that is either a West Bank occupied and progressively settled by Israel or as providing some form of nuclear area for a Palestinian entity. However, we will see that to choose to proceed towards continuing Israeli occupation and progressive settlement does not remove the 'Palestinian threat'. Indeed such an expansionary aim has quite specific and potentially profound ramifications upon the adjacent Arab states, including those so-called conservative Arab states such as Saudi Arabia and Jordan that have indicated a willingness to recognise Israeli's right to exist. The Fahd Plan implicitly suggested precisely such a recognition, envisaging a Palestinian political entity in the West Bank and the Gaza Strip and calling for a broader discussion than the Camp David peace process.

Included in the Saudi proposal are a withdrawal of Israel from areas occupied since 1967 and removal of all Israeli settlements from those areas; guarantees of freedom of worship for all; recognition of the rights of the Palestinian refugees from the 1948 and 1967 wars, their repatriation and compensation; transitional U.N. trusteeship over the West Bank and Gaza; the establishment of an independent Palestinian state with its capital in East Jerusalem; an assurance for the right of all states in the region to live in peace - all this to be guaranteed through the U.N.

Of course there is much in such a plan that is unpalatable to both Israel, the adjacent Arab states and the Palestinians, and the Plan was firmly rejected at the 12th Arab Summit held in Fez, Morocco in November 1981. Yet such proposals, contained within the Fahd Plan, may be seen as bases from which to proceed toward Solomonian compromises, perhaps to be considered by, on the one hand, a group of moderate Arab states including Egypt, Jordan, Lebanon, and Saudi Arabia, and on the other Israel, overseen by the U.S., the U.N. and the E.E.C.

As political geographers it is well to recognise our particularly spatial and 'real-world' viewpoints in power conflicts. Deriving from this spatial view certain matters will be developed. Whereas Johnston seems to argue vociferously regarding the lack of advancement in political geography relating to the 'people-territory-state' linkage (Johnston 1981: 19) it is suggested here that, political scientists have generally failed to incorporate real spatial inputs in any satisfactory manner into their conceptualisations. In essence the dispersal of much of the Palestinian Arab population into neighbouring Arab states is the basic reality from which to proceed. Furthermore, from the war of 1967 there emerged certain ominous signs within the Palestinian Liberation Movement which supposedly possess 'revolutionary' attitudes (Sayigh 1979).

The various factions which comprise the P.L.O. are united in their fidelity to the land, by their hope for eventual return to the land and through their armed struggle. By contrast, it is suggested, the development of a revolutionary spirit would, in part at least, seek to re-evaluate the fundamental nature and structure of society, both as it exists and potentials for change, and to consider social and capital formations.

The romantic view whereby the downtrodden 'seethe with revolt' and move to a revolutionary posture is largely dismissed by Davies (1962) who, in his considerations of the conditions likely to lead on to revolutions, suggests that often such folk are overwhelmingly pre-occupied with just staying alive or working towards a real, immediate, tangible objective. For the Palestinians this opportunity would be the realisation of a national homeland and not necessarily a commitment to a revolutionary ideology. Revolution, in the sense of an attempt to alter society radically by political means, depends upon the development of such a commitment. A particular feature of the Palestinian problem is that, in time, without achieving a solution generally acceptable to the Palestinian people, it is believed that a socialist-revolutionary approach would become increasingly attractive. In such a development towards a revolutionary spirit the Palestinian problem should be set in a broader context, perhaps of the entire Arab world. In essence it is being suggested here that the Palestinians are working within a diaspora that contrasts with the usual basis of the spatial political state. The bounds of this realm extend far beyond the confines of Palestine.

The Palestinian thinking could well proceed on from a national liberation movement to a fundamental appreciation of Israel in its imperialistic setting and to question the order that gave it birth. Indeed for the development of socialism it might be suggested that:

'To travel hopefully is a better thing to
arrive, and the true success is to labour'
(R. L. Stevenson, 1891, Ch. 6).

It is 'environmental pressure', by and large the output function
of the state, that literally brings the system down to earth.
By contrast the Palestinian stays 'in the air' with an increasing
likelihood of a revolutionary effect to the entire Arab 'nation'
and ultimately to Israel. Thus, whether a Palestinian state is
in fact realised or whether a revolutionary fervour develops,
fundamental changes will occur throughout the region.

SUMMARY AND CONCLUSIONS

In this essay we have briefly considered the notion of
Eretz Yisrael, paying particular attention to the West Bank and
to certain of the plans for its settlement and establishment of
'new facts'. I have sought to demonstrate that the choice is not
between an Israeli occupation of the West Bank (stability) or
some form of a Palestinian political entity (instability). Either
way the signs are ominous and compromises must be actively sought
to lessen the tensions. The alternative is that mounting
Palestinian revolutionary fervour could have a potential knock-on
effect to the so-called conservative Arab states. Indeed such
states have as much, if not more, to fear, as does Israel, from
a far-leftist oriented Palestinian revolution. The alternative
to reason and compromise appears to be for further and mounting
crisis which would lead inexorably to war on an ever- enlarging
scale.
Finally the marked increases in the size and density of the
refugee population and its developing political consciousness
augurs a continuing crisis within host communities and recurrent
outbreaks of violence. It is my firm belief that there can be
no reductions of tensions nor increased stability within the
Middle East until the refugee problems are faced squarely and a
just, equitable and lasting solution obtained for all parties.

Acknowledgements   I take this opportunity to thank Nurit
Kliot and Stanley Waterman for their assistance and friendship,
and the insightful comments of Saul Cohen and an unknown reader
in enabling me to update this paper.

REFERENCES

ARDREY, R. [1966]
   The Territorial Imperative, (Atheneum, New York).
BRUEGGEMANN, W. [1977]
   The Land: Place as Gift, Promise and Challenge, (Fortress
   Press, Philadelphia).
CALHOUN, J.B. [1971]
   'Space and strategy in life', in A.H. Esser (ed.), Behavior
   and Environment, (Plenum Press, New York), 329-87.
DAICHES, D. [1975]
   Moses, (Weidenfeld and Nicolson, London).
DAVIES, J.C. [1962]
   'Toward a theory of revolution', American Sociological
   Review, 27, 5-19.
DIVREI HAKNESSET [1949]
   Quoted from M. Brecher (1977), 'Jerusalem: Israel's political
   decisions, 1949-1977', The Middle East Journal, 32, 13-
   45.
DROBLES, M. [1978]
   Master Plan for the Development of Settlement in Judea and
   Samaria, (World Zionist Organisation, Jerusalem).
DROBLES, M. [1980]
   Strategy, Policy and Plans for Settlement in Judea and
   Samaria, (World Zionist Organisation, Jerusalem)
EICHRODT, W. [1961]
   Theology and the Old Testament, (The Westminster Press,
   Philadelphia).
ESSER, A.H. [1971]
   'The importance of defining of spatial behavior parameters',
   in A.H. Esser (ed.), Behavior and Environment, 1-8.
FISCHER, F. [1971]
   'Ten phases of the animal path: behavior in familiar
   situations', in A.H. Esser, (ed.), Behavior and Environment,
   9-21.
HALL, E.T. [1966]
   The Hidden Dimension, (Doubleday, New York).
HEDIGER, H. [1949]
   'Saugetier - Territorien and ihre Markierung', Bijdragen
   tot der Dierkunde, 29, 172-84.
ISAAC, R. [1976]
   Israel Divided: Ideological Politics in the Jewish State,
   (Johns Hopkins University Press, Baltimore).
JACOBSON, D. [1982]
   The Story of Stories: The Chosen People and its God, (Secker
   and Warburg, London).
JOHNSTON, R.J. [1981]
   'British political geography since Mackinder: a critical
   review', in A.D. Burnett and P.J. Taylor (eds.), Political
   Studies from Spatial Perspectives, (Wiley, Chichester),
   11-31.

KARMON, Y. [1971]
Israel: A Regional Geography, (Wiley, New York).
LEYHAUSEN, P. [1971]
'Dominance and territoriality as complemented in mammalian social structure', in A.H. Esser (ed.), Behavior and Environment, 64-81.
LORENZ, K. [1967]
'The evolution of behavior', in Psychobiology: Readings from Scientific American, (Freeman, San Francisco), Part 1, Chapter 5.
NEWMAN, M.L. [1962]
The People of the Covenant: A Study of Israel from Moses to the Monarchy, (Abingdon Press, New York).
NISAN, M. [1978]
Israel and the Territories: A Study in Control 1967- 1977, (Turtledove Publishing, Ramat Gan).
ROWLEY, G. [1981]
'The land in Israel', in A.D. Burnett and P.J. Taylor (eds.), Political Studies, 443-65.
ROWLEY, G. [1983]
'Israel and its land: A Reconstructuralist approach', (in preparation).
SAYIGH, R. [1979]
Palestinians: From Peasants to Revolutionaries, (Zed Press, London).
SMITH, J.Z. [1969]
'Earth and the Gods', Journal of Religions, 49, 107-21.
STEVENSON, R.L. [1888]
Virginibus Puerisque, (Nelson, Edinburgh).
TUAN, Y.F. [1977]
Space and Place, (University of Minnesota Press, Minneapolis).
TUAN, Y.F. [1979]
Landscapes of Fear, (Blackwell, Oxford).
UNITED NATIONS [1976]
General Assembly Records: Thirty-First Session, Supplement 13, (New York).
WYNNE-EDWARDS, V.C. [1962]
Animal Dispersion in Relation to Social Behaviour, (Oliver and Boyd, London).

PART IV: PLURALISM —
THE STATE AND THE WORLD

Chapter 16
ON THE INSTRUMENTAL FUNCTION OF POLITICAL GEOGRAPHY

Shalom Reichman

INTRODUCTION

In this paper, several basic functions that geographical space can fulfil in the realm of government actions are outlined, demonstrating the rationale of such functions on the grounds of either efficiency or equity criteria. In addition, some typical problems encountered in the course of the application of policies which have a significant spatial component are examined. The 'instrumental' function of political geography is defined as the subject of scientific inquiry which examines the intentional use of space by government in the furtherance of its political goals. Though related to the 'spatial theme' in political geography (Kliot 1982), the instrumental approach differs from the main body of research in the field, viewing issues from the perspective of the political actor. As such, it is probably closer to the spirit of recent work in the geography of public finance (Bennett 1980: 28-32, 61-72; Johnston 1980).

That space can be used for the attainment of political objectives is as old as history (Gottmann 1966). There may be, however, considerable variation in the potential contribution of geographical factors in achieving the basic objectives that usually underly state interventions. These are (a) the provision of security and order, (b) the stimulation of economic growth and well-being and (c) the guarantee of social justice.

Thus, it has been long recognised that geography plays an important strategic role in the defence of a territory, illustrated by the fact that spatial concepts, such as <u>heartland</u> and <u>rimland</u> were rapidly assimilated into the terminology of political geography and related disciplines.

The importance of geography as means to attain economic goals is probably less than that for security needs. The concept of rent, as noted by the early political economists, is a good example of a potential geographical effect. Adam Smith recognised this when he lauded the benefits that accrue from 'breaking down the monopoly of the country' by the provision of a transportation infrastructure (Smith 1937 edition).

With respect to the third objective, assuring social justice, geography appears as a relatively minor factor. For example, in the liberal theory of justice expounded by Rawls (1971), spatial factors are absent from the definition and discussion of both procedural and distributive justice. Actually, as more programmes are targeted at specific populations characterised by their places of residence, it is clear that spatial factors will gain in importance as contributors to the provision of primary social needs.

GEOGRAPHY AS AN INSTRUMENT OF STATE INTERVENTION

The potential contribution of geographical elements to various types of state interventions is directly related to the fundamental properties of space. At the most basic level, that of the individual actor, these properties can be formulated in terms of the duality of space as an opportunity and constraint. Thus, for example, activities of an individual are bounded. He may take advantage of numerous resources and services in a given radius subject to the constraint of rising costs due to increasing distance.

At the aggregate level of decision-making, that of government, space retains the same duality, though it is more general. It provides the resource base for strategic, economic and welfare purposes and, at the same time, exposes the user to natural and man made impediments which act as barriers. In the French geographical tradition, these properties are summarised by the concepts 'espace-support' and 'espace-obstacle' (Claval 1978: 15-17). The 'supporting' property consists of the provision and exchange of materials, goods and services, whereas the obstacle property refers to the existence of frictions due to distance, as well as to various natural and institutional barriers.

Governments have the option of reinforcing either of these two properties of space. They may decide to enhance the 'fluidity' or transparency of space by increasing accessibility and by removing natural or institutional barriers, facilitating free circulation of people, goods and information. On the other hand, space can be made more impenetrable or opaque by raising barriers so as to impede the movements of an invading force, for example. This is frequently the main argument in favour of 'secure' boundaries. Similarly, a government may choose to curb the dissemination of knowledge, or to prevent the free circulation of people within its jurisdiction.

In practice, space is manipulated subtly in order to achieve various political objectives. Notwithstanding this subtlety, the outcome of such manipulations should result in a dichotomy between (a) outcomes which make space more discontinuous, by establishing controls and limits and (b) outcomes that enhance space as an open system, stressing its generic continuity by removing barriers and reducing friction.

The dichotomy in the use of space for political ends need not be attributed solely to ideological divergences among governments or among social classes. While it is true that utilitarianism and protectionism differ widely in the posture of the government towards space, ideological factors are not a necessary condition for the appearance of the contrasting outcomes described above. On the contrary, a government with a given ideology may find itself in some cases in a situation where it is called upon to 'open up' space, and in other cases, to erect barriers.

What rationale, other than ideology or self-serving partisan ends, underlies the manipulation of space by governments? To answer this, we can turn to other disciplines. Economics, for example, has raised this issue in general terms by questioning the justification for government intervention at all. Interestingly, when economists examined this problem, they often chose to use spatial illustrations. This selection of geographic examples is not accidental and reflects the unique property of space as being continuous. Consequently, when land is a factor in an economic process, we find numerous instances of positive and negative spillovers or externalities.

In such cases, the rationale for public intervention would be to internalise these externalities by bringing them into the price mechanism, either directly or indirectly. The method used to achieve this result is precisely that of government intervention, with the aim of modifying space by making it either more or less contiguous or 'open' than before, with a resultant increase in efficiency or equity of some economic process occurring in a given area.

Tullock (1970) provides us with an example of how the market mechanism fails because of a positive spillover effect of 'geographical contiguity'. Under current technology, aerial spraying cannot be prevented from spilling over the property lines of a single lot. The owner of the neighbouring lot, who benefits from the spillover has no incentive to share in the costs of the service as he gets it free. As a result, the owner of the lot who wants to purchase the spraying service pays an unjustfiably high price and may well forego this altogether with concomitant market failure. However, if local government steps in, imposing a collective charge for spraying, the cost borne by each individual is smaller than that paid by each purchaser in the free market. Consequently, everyone is better off, the service being provided at the lowest possible cost (Figure 16.1).

Three major conclusions can be drawn from this example. First, government intervention disregards existing limits (the grid of individual properties). The removal of this 'stickiness' results in a more 'open' space. Second, the application of efficiency criteria generally leads to the formation of larger territorial units in which certain actions occur (from individual lots to the whole territory). Third, even with larger territorial

units, the phenomenon of spatial spillovers does not vanish
(Figure 16.1), although its areal extension is considerably
reduced.

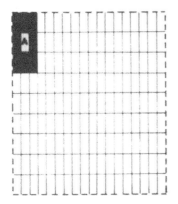 

----   Limits of incorporated
        area

A   Area of Individual
        property

        Minimal zone of aeral
        spraying

        Properties benefiting
        from free spraying=
        positive spillovers

Figure 16.1: Mosquito Spraying Model with Market Failure (left)
and with Collective Imposition of Spraying Fee (right).

It should be noted that several problems are not discussed
in this example, including the optimal territorial size for a
given public service, the solution to which requires real cost
data for a given service in a specific area and exceeds the scope
of this paper.
Another example of spatial manipulation is provided by
Hotelling's (1929) ice-cream vendor model modified to allow for
political intervention (Hirschman 1970: 66–68). In the unconstrained
market model, stability in the locational competition of two
ice-cream vendors is achieved when they are located side by side
at the centre of a linear market. However, from the clients'

point of view this location is not equitable as satisfaction of the clients is affected by the cost and effort of overcoming distance to the vendor. This spatial factor, tapering, underlines the erosion in the utility of a given service as a function of the distance to the clients. The tapering in this example is not linear, but increases beyond a certain distance as the movement is by pedestrians.

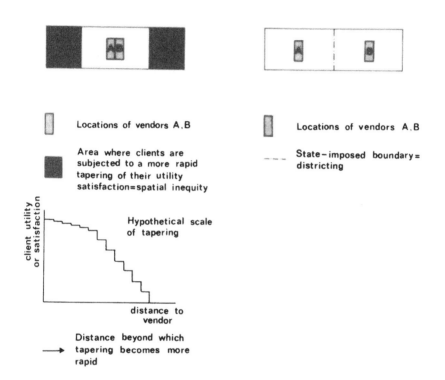

Figure 16.2: Ice-cream Vendor Model Without Institutional Intervention (left) and With State-Imposed Districting (right).

Two groups of clients can be identified. Some are located in areas adjacent to the vendors and experience a low rate of tapering; others are located further away and suffer from a high rate of tapering and are hence subjected to a typical form of spatial inequity.

The government intervention in this case involves districting, the setting up of a boundary dividing the service area into two parts. Each district has a single vendor who serves only those clients in his area. As a result of districting, spatial decentralisation occurs following the relocation of the vendors at the centre of their respective monopolistic service areas. The clients no longer suffer from spatial inequity since the maximum distance to the vendor remains below the range where the tapering is more rapid (Figure 16.2).

He~e again we can draw some general conclusions about the spatial manipulations. First, the intervention involves the erection of a boundary that previously did not exist so that the space becomes more confined and controllable. Second, the application of equity criteria, with its orientation towards the clients, will generally result in the creation of smaller territorial units. Finally, even in smaller territories, the phenomenon of spatial tapering does not disappear but its acuity will obviously diminish

Limiting ourselves to the issue of minimisation of delivery costs only, decentralisation represents a compromise between efficiency and equity. But when other factors, such as market size and scale economies in the level of service, are included, decentralisation may lead to inequities in the quality of service and substitutes one negative externality with another (Tapiero & Desplas 1979).

From a methodological viewpoint, another conclusion seems appropriate. From a given state of spatial openness, government may need to and can modify it in opposite directions. This should lead to the introduction to political geography of methods of analysis particularly suited to the study of such potential bifurcations. These methods refer to concepts and structures derived from catastrophe theory (Thom 1975; Zeeman 1977) which focuses on problems associated with discontinuities, their causality and on the conditions of stability and instability engendered by different forms of discontinuities. These approaches have only recently begun to be integrated in policy-oriented geographical literature (Dendrinos 1980; Massam 1980). It is suggested that the instrumental function of geography, as presented above, lends itself, in principle, to a promising qualitative analysis in terms of catastrophe theory.

RECURRE:T PROBLEMS IN THE APPLICATION OF POLICIES WITH SIGNIFICANT SPATIAL CONPONENTS

Another approach to the study of the instrumental function of political geography is by drawing generalisations from the observations of everyday practices, rather than deducing them from theoretical concepts (Reichman 1978). Governments or public agencies carry out a wide range of policies and programmes which contain significant territorial components. Some explicit examples involve objectives such as colonisation, industrial development and population dispersal or programmes containing some form of resource distribution aimed at specific urban neighbourhoods. At the other end of the scale are policies and programmes with implicit geographical components including routine government agencies with a line function such as the delivery of services or the maintenance of administrative control over a given territory. Two questions can be asked: (a) Are there any recurrent problems in the real world application of such programmes and policies and (b) if so, what can geographers suggest in the way of alleviating these problems?

Problems caused by geographical components of policies and programmes are not necessarily recognised as such by the agencies in charge of their implementation. Problems are often identified in management terms, within the framework of organisational behaviour and structure. The following review, based on participant-observer experience, presents the problems in both geographical and management terms. The three most frequent are listed in Table 16.1.

TABLE 16.1: FREQUENT PROBLEMS ENCOUNTERED IN POLICIES
AND PROGRAMMES WITH SIGNIFICANT GEOGRAPHICAL
COMPONENTS

	Type of problem in geographical terms	Type of problem in management terms
1.	Spatial reach of authority	Spans of control, decentralisation
2.	Modifications or congruence of boundaries and limits	Inter-agency or lateral co-ordination
3.	Time span of geographical processes and patterns	Programme reformulation or termination

Spatial reach   Many agencies with a line function have an organisational structure with territorial components which can be represented on a map. This structure is usually hierarchical with nested levels, such as the division of a country into national, regional and local administrations or of a region into metropolitan areas, cities and boroughs. An important problem in such a structure is that of the spatial delegation of authority to the various levels and its actual reach, where authority consists of making rules and regulations and of financing the activities.

Typically, this is difficult to solve, because of the existence of geographical and organisational contradictions and biases which preclude the establishment of simple rules as to when centralisation is to be preferred to decentralisation and vice-versa. On the geographical level, proximity enables a better understanding of local needs, enhancing equity. But, at the same time, the creation of local power foci may result in the breaking up of spatial continuity, affecting efficiency negatively. Similar observations apply to organisational structure. Decentralisation is associated with greater democracy, better first-hand information and faster decision-making; centralisation has clear advantages in terms of co- ordination, efficiency and public control.

In addition, several biases are introduced by organisational behaviour. For example, Morris (1968) hypothesised that geographical decentralisation is seldom followed by decentralisation of communications and control; on the other hand, decentralisation of communications and control is often followed by geographical decentralisation. Another example of asymmetry applies to behaviour in the hierarchy, it being suggested that each level in the organisational hierarchy seeks to be more decentralised with respect to its superior, while striving to maintain or increase the degree of centralisation vis-a-vis its subordinates (Mackenzie 1978).

Geographers can offer the well-developed concepts of threshold and range which are operational in terms of the delivery of government services. These methods may be usefully complemented by network analysis, where authority can be represented by nodes and flows of information and decisions. Also, by assuming that some links in the hierarchical network are biased, some of the biases noted above are dealt with. The degree of autonomy of the agency as a whole may be determined by metwork indices which may allow us to identify whether an agency has a market or resource orientation, where resources are defined by proximity to the source of authority.

The outcome of such analyses can be used to describe the spatial reach of a given agency by weighting the number of functions (e.g. financial, regulatory, certification, staffing, advisory) over which each node has control. They can also focus on the node or level where there is an accumulation of authority and relate it to its intended spatial reach so that discrepancies might be attributed to factors either within or outside the organisation. These analyses could also identify loose or 'neutral' agencies and functions with relatively few interactions with subordinate or superior levels in the spatial hierarchy. Such agencies are known to be good candidates for relocation in places other than the administrative centre in countries with a strong spatial dispersion policy (Lavigne 1976).

Modification and congruence of boundaries and limits

The delimitation of boundaries and administrative limits is one of the classical tasks of political geography and in many countries, geographers have been actively engaged in performing these tasks. What makes boundary modification so complex? From the perspective of organisational structure and behaviour, we can identify three main contributing factors. First, we must recognise the context in which the process takes place. Many state and local administrations view space in a game- theoretic framework, assuming the property of zero-sum in terms of exclusive control. From there is but one step to the tendency to equate control over territory with power; consequently, any shift in boundaries will be regarded as strengthening one party and weakening the other. However, this tendency is not uniform and becomes most evident in strong, centralised structures such as military authorities which are generally far more sensitive to the spatial limits of their jurisdictions than are civilian authorities.

The complexity is also related to the substance and form of the geographical contribution to boundary modifications. The substance of the geographical assessment is composed not only of quantifiable factors such as resources, population groups and circulation, but also of qualitative factors such as territorial belonging and attachment. Consequently, when alternative delimitations are formulated in terms of territorial 'price-lists', there is a serious problem of ordering and weighting both quantitative and qualitative factors in the same 'price-list' (Reichman 1982). As to form, the spatial assessment needs to be presented so that it can be incorporated with non- spatial arguments based upon political, administrative, public opinion and other considerations which affect the decision.

The process of resolving the power conflict implied in boundary modifications should also be mentioned. Rather than being resolved at the local level, which is generally the level most affected, negotiations over boundary modification tend to creep upward in the political and administrative hierarchy. Eventually, conflicts over administrative limits are adjudicated by one of the branches of central government which is accepted as 'common ancestor' by all parties. In the case of international disputes, process factors become even more important, but this issue canot be treated here.

The problem of lack of congruence of different administrative limits in the same territory is as common as that of boundary modification. This may result from the different functions of government agencies, but the more interesting cases are those based on the mismatching of administrative limits which result from strong historical and cultural influences, and the territorial limits of new innovative programmes designed to serve areas or populations on the basis of different spatial criteria.

From an organisational viewpoint, the problem lies in maintaining administrative efficiency and control despite the existence of distinctive spatial limits in the same territory. From a geographical perspective, it is perhaps just as important to assess whether this coexistence of non-congruent limits can be maintained over time. It is difficult to draw general rules for this kind of situation since it depends upon many factors outside the scope of geography. Nevertheless, it can be observed that the more entrenched the agency by means of statutory or traditional power, the more likely it is to cause eventually the congruence of the spatial limits of another, less powerful agency with its own. When both agencies have enough power to withstand the pressure of congruence, the problem of coexistence becomes one of coordination. Whether such coordination is really occurring can be determined from the existence of relevant statistics. Thus, for example, spatial data for each of the different jurisdictions that can be amalgamated are possible indicators of coordination. This is particularly true for the geographical distribution of finances in general and the national budget in particular.

Time-span and geographical processes

This issue is probably the most difficult of all, since it involves a structural contradiction between geographical perspective, based on objective evaluation and prediction of processes, and the political viewpoint, which attempts to achieve certain objectives, some of them within a specific time framework. As an example, let us consider two typical decisions with a geographical component: (a) the establishment of an authority to develop a remote region, by investments in transport facilities, in economic opportunities and in new settlements and (b) the setting up of an agency designed to administer a programme of housing rehabilitation in depressed urban neighbourhoods. For each of these projects, geographers may be able to assess the time needed to reach a set of spatial targets based on resources available and on first-hand knowledge of conditions in the field. These natural and man-made conditions may frequently act so as to inhibit the tempo of the spatial processes initiated by such agencies.

The time horizon of agencies in charge of policies with clear political pay-offs does not generally coincide with that recommended by professional assessments of social scientists. On the contrary, it is frequently influenced by the need of public approval and legitimation. In other words, objectives are defined in terms of expediency and practicability from a political and organisational viewpoint. Consequently, in many cases targets are determined in such a way that they can be reached within shorter time spans than the normal development of changes in spatial structures.

On the other hand, the logic of organisational behaviour has as a prime objective the survival of the organisation. We can frequently observe that agencies are perpetuated beyond the period needed for the fulfilment of their task (Kaufman 1976). This brings us to the potential contribution of geographers to the issue of policy termination (Bardach 1976; Behn 1978). It is expected that in this problem area we will find fundamental differences in the opinions of the three main groups of actors - politicians, civil servants and professional advisers. The political actor is generally more interested in short-term results. Professional advisers, including geographers, usually recommend longer, but finite time-spans for the implementaton of a spatial programme. In between are the civil servants who must accept the directives of the political level of government but at the same time appreciate the relevance of suggestions put forward by professionals. Notwithstanding these conflicting pressures, civil servants also have their own interest, namely the survival of the organisation. Given this complex behavioural situation in the organisation, it is difficult to predict whether the political or the professional outlook towards the time limits of a spatial programme will have the upper hand.

CONCLUSION

This brief overview of some of the theoretical concepts and empirical generalisations of the potential use of geography in the service of government may be summed up in several general statements.

First, as a concept, the spatial dimension offers itself to a manipulation in two opposite directions - reinforcing the property of space as an agent supporting human activity, or alternatively intensifying the property of space as a barrier to human activity. We may add that in the perspective of the French geographical paradigm of progress, whereby Mankind is in the process of freeing itself from the bonds of Nature, it is possible to attribute value-loaded terms such as positive and negative to the alternative terms of intervention.

Second, in the empirical running of agencies charged with the implementation of policies with a spatial component, it is possible to identify a number of problems related to these properties of space. One way to overcome such problems is to specify spatial elements so as to devise trade-offs with non-spatial factors, embedded in organisational structures and behaviour. This option does not work whenever the issue at stake is connected to time-spans within which a programme has to be terminated.

Finally, these observations might be useful didactically in developing both curricula and practical methods to assist the 'instrumental' function of political geography in serving the needs of government agencies. In terms of curricula, it is suggested that regional textbooks address themselves to the development of the political and administrative landscape just as natural features such as landforms and climate are currently discussed. More generally, the notion of a 'geographical niche' in the field of public policy studies should be recognised and encouraged. This will lead, hopefully, to a greater openness in the social sciences.

REFERENCES

BARDACH, E.C. [1976]
'Policy termination as a political process', Policy Sciences, 7(2), 123-32.
BEHN, R.D. [1978]
'How to terminate a public policy: a dozen hints for the would-be terminator', Policy Analysis, 4(3), 393-413.
BENNETT, R.J. [1980]
The Geography of Public Finance, (Methuen, London).
CLAVAL, P. [1978]
Espace et Pouvoir, (Presses Universitaires de France, Paris).
DENDRINOS, D.S. [1980]
Catastrophe Theory in Urban and Transport Analysis: eight papers, (Research and Special Program Administration, Department of Transportation, Washington D.C.).
GOTTMANN, J. [1966]
'Geographie Politique' in Geographie Generale, (Encyclopedie de la Pleiade, Gallimard, Paris), 1749-65.
HIRSCHMAN, A.O. [1970]
Exit, Voice and Loyalty, (Harvard University Press, Cambridge).
HOTELLING, H. [1929]
'Stability in Competition', Economic Journal, 39, 41-57.
JOHNSTON, R.J. [1980]
The Geography of Federal Spending in the U.S.A., (Research Studies Press, Wiley, New York).
KAUFMAN, H. [1976]
Are Government Organizations Immortal?, (The Brookings Institution, Washington D.C.).
KLIOT, N. [1982]
'Recent themes in political geography: a review', Tijdschrift voor Economische en Sociale Geografie, 73(5), 270-79.
LAVIGNE, P. (ed.) [1976]
Dispersion Geographique des Administrations Centrales de l'Etat, (Economica, Paris).
MACKENZIE, K.D. [1978]
Organizational Structures, (AHM Publishing Corporation, Arlington Heights).
MASSAM, B.H. [1980]
Spatial Search, (Pergamon Press, Oxford).
MORRIS, W.T. [1968]
Decentralization in Management Systems, (Ohio State Press, Columbus).

RAWLS, J. [1971]

A Theory of Justice, (Harvard University Press, Cambridge).

REICHMAN, S. [1978]

'Le voyage d'un geographe dans le "pays de l'administration"', Espace Geographique, 7, 123-25.

REICHMAN, S. [1982]

Factors in the Delimitation of the Eastern Boundary of the State of Israel, (The L. Davis Institute for International Relations, Policy-oriented paper #5, The Hebrew University of Jerusalem), (in Hebrew).

SMITH, A. [1776, 1937]

The Wealth of Nations, (Reprinted in The Modern Library, Random House, New York).

TAPIERO C.S. and M. DESPLAS [1979]

'Location with an equity criterion', in J. Beaujeu- Garnier and S. Reichman (eds.), Urbanisation Contemporaine et Justice Sociale, (National Council for Research and Development, Jerusalem), 276-95.

THOM, R. [1975]

Structural Stability and Morphogenesis, (W.A. Benjamin, Reading, MA).

TULLOCK, G. [1970]

Private Wants, Public Means, (Basic Books, New York).

ZEEMAN, E.C. [1977]

Catastrophe Theory, Selected Papers 1972-1977, (Addison-Wesley, Reading, MA).

Chapter 17
THE TRANSFORMATION OF LOCAL GOVERNMENT FROM COUNTIES TO REGIONS

Andrew F. Burghardt

The past two decades have seen attempts in several countries to reform the system of local government and administration. The old system had been found wanting not only for its distribution of functions and responsibilities but also for its geographic structure - the shapes and sizes of its territorial units and the existence of internal separators. The reform of local government is thus eminently a geographical problem even though its articulation, solution and implementation have fallen mostly into the hands of economists, lawyers and politicians. This work deals primarily with the experience of such reform in Ontario, Canada.

The old system is still familiar to much of the Western World. The province or state area is divided into counties or equivalent units and those intermediate level units are subdivided into local units such as townships. Incorporated areas are seen as exceptions and are excluded from the administration of the county. These features vary from place to place and are more complex than stated here, but in general, the city and county run their affairs separately from one another. The city limits are barriers which require agreements across or directives from above in order to traverse. It is this separation of city and surrounding areas which has caused considerable problems of administration and planning. In Ontario the county-township system was established soon after the American Revolution and was given its definitive codification in 1850.

Complaints against the old system form a list familiar to many geographers. Improvements in transportation, particularly the widespread use of the automobile have increased personal mobility and in industry and commerce. This has made nonsense of the old city limits. Fortunately, the American experience of an explosive growth, leaving a fiscal vacuum near the old centre, has not yet been widely duplicated elsewhere. Long-distance and cross commuting are common as is the phenomenon of many families who opt to live on some country road while maintaining most of their other ties with a nearby city. Moreover, attempts to move the city limits outward have met with such strong resistance that even those governments which do have the power to decide such matters such as Ontario, have shied away from annexation.

Our era has also witnessed rising demands for services with the belief that such services should be applied equally across the territory and across the population. Such demands have placed enormous strains on small political units with poor financial resources and have also deprived them of the opportunity of finding, hiring and maintaining good professional staff. An inevitable, fierce competition for industrial and commercial assessment has resulted, which has made planning extremely difficult. Sorely beset, the Ontario municipalities have turned to the body which holds ultimate responsibility, the provincial government. Swamped by the cries of distress and pleas for supplementary funds from some one thousand units of local government, the Ontario government cannot be blamed for envisaging an administrative Eden where the province would consist of vastly fewer units, each of which would set and keep its own house in order.

The higher levels of government have accepted a number of basic ideas and evaluations, foremost among which has been the primacy of Economics and Management (Smith Report 1967; Stewart Report 1978: 2). This was probably inevitable, considering the financial dimensions of the problems. Economics is now, without doubt, supreme among the Social Sciences in the opinion of decision-makers. Efficiency, perhaps the most revered goal of our times, is usually expressed in economic terms.

In contrast to the rise of Economics to preeminence has been the relative decline of History. History is still accepted as an invaluable guide to the development of the self- knowledge of a people and as an indicator of how we have come to our present situation, but it has clearly lost the role it once had of suggesting how we should confront the future. However, historical associations and artifacts have become increasingly meaningful to the inhabitants of many areas. The emphasis on the self-identity of small places and the rise of the preservationist movement have led many people to place History above Economics. When these two viewpoints clash, as they often do in local government reform, positions become ever more entrenched, especially the positions of those articulate persons who feel that they are defending old values against a commercial or governmental juggernaut.

One feature that has come out of the past and which cannot easily be ignored is the network of boundaries which have framed the counties and townships. Planners as well as local inhabitants have perceived space not as a boundless tabula rasa, but rather as broken up into sets of discrete units which are nested within each other and which are to be the territorial building blocks of the new order.

The emphases on efficiency, on good management and on regional planning have worked towards the creation of larger units. Many planners believe that a city region should be planned in total and know from experience that the fewer political jurisdictions one must work with or satisfy, the more quickly and better the task can be completed. Unfortunately, the creation of larger units has the tendency to move government away from the people. Even if the representative ratio will not be excessively high, the very scale of the new structure makes it more intimidating and hence more remote (Spectator 28 May 1981: 10).

Three approaches have been utilised in recent reorganisations. These may be labelled the traditionalist model, the homogeneous region model and the nodal region model. Because they seem to typify the approaches taken in those countries, they may also be named the American, English and Scottish models respectively, although they have also been applied elsewhere. Each of the three has a clearly recognised top priority.

In the traditionalist or American model, the areas and boundaries of the units concerned are maintained. Local democracy is held to be preeminent and territorial changes occur only when the local populace so decides. Under this constraint, annexations and amalgamations are rare and reform is commonly achieved through efforts at cooperation and through selective funding. Problems of servicing are met by the establishment of a plethora of ad hoc single-function bodies. Since this approach allows for the continued existence of small units and seems to promise generous funding, it is the approach most favoured by local officials everywhere.

The homogeneous region or English model is a variant of the traditionalist model in that it maintains the administrative contrast between rural and large-scale urban. More than that, it can be said to increase the contrast. New boundaries are drawn to encompass all the urbanised areas of a large city or conurbation, plus a narrow fringe to allow for future expansion. In a sense, this is a massive annexation. The effect is to separate the giant urban complexes from the rural areas and their central places, leaving intact as many of the old counties as is possible. Thus, for example, in the United Kingdom, Greater Manchester and Merseyside have been cut way from Lancashire and Cheshire; West Yorkshire and South Yorkshire from the remainder of Yorkshire; Tyne and Wear from Northumberland and Durham. The English veneration of an idealised countryside may form one of the bases of this approach. The 'dark satanic mills' are to be kept apart from 'England's green and pleasant land' (Blake in Erdman 1965: 95).

The nodal region or Scottish model is the ideal of most regional planners. City regions (Dickinson 1964; 1967; 1970), or the central cities and their immediate hinterlands (umlands), are delimited and become the new administrative units in addition to their role as planning units (Senior 1965; 1966; Robson 1968; James, House and Hall 1970). It is essential that every region have a focal centre, preferably one dominant central city. Rural areas are seen not in their own terms, as in the first two models,

216

but as subsidiary parts of a functional econo-administrative system, the locus of power of which is at the focus of a structural web. Regions thus vary in size according to the population and the attractive powers of the focal centre. Hence, in Scotland, half the population was placed in a single administrative region based on Glasgow, which was extended to include several of the remote islands of the Inner Hebrides.

In the case of Ontario, the first major break out of the county-township system occurred in the 1950s with the creation of Metropolitan Toronto, a union of the central city and its suburbs. The delimitation of the area was not precisely on the basis of functional ties, nor even on planning needs, but rather on the existence of several townships at the southern end of York County, which constituted most of the contiguous built- up area. Thus, it most closely approximated the homogeneous region or English model. A two-tier or miniature federal system was established to allow for administration at both the regional and local municipal levels.

Metro Toronto has been a pilot case for the reorganisation of Ontario and the two-tier system pioneered there has been applied in the subsequent regions. However, it must be borne in mind that in both scale and type, Toronto is unlike all the other Regional Municipalities in Ontario. The entire area is urbanised so that it is, in essence, one city, which is now hemmed in by straight-line limits which separate it from the true suburbs outside. The organisation of the new suburbs as units distinct from Toronto has signalled a gradual drift towards the American model.

The first step towards reform outside Toronto was the governmentally sponsored Smith Report on Taxation of 1967. As the title suggests, this was an economic analysis, and rather predictably called for the amalgamation of the many existing units into fewer, larger and more viable units. On the basis of the Smith recommendations, the Ontario Government moved to transform all of Southern Ontario and the urban areas of Northern Ontario into a set of viable administrative areas to be called Regional Municipalities. The duplication of facilities among competing small municipalities would be eradicated. This and other anticipated savings would provide welcome reductions for the local taxpayers, tangible proof that the government had been right.

There was, however, a fundamental difference between the way Ontario initiated the reform and the way it was undertaken in Great Britain and elsewhere. In each of the British cases the entire country was reorganised as a result of one commission, one process, one Act. In Ontario the government chose to commence the process on a piecemeal basis. Individual commissions were established for each of the perceived 'Regions' of the province; no overall plan for all of Southern Ontario was to be attempted. Perhaps because each of the commissions was expensive and required highly-skilled professionals over fairly long periods of time, the commissions were established for only a few areas, principally in the urban areas west of Toronto.

It is not clear why the provincial government opted for the piecemeal approach which, with its multiplication of commissions, was certain to be expensive, rather than attempt one overall restructuring. An attempt to do so would probably have seemed too overwhelming a task, and in all probability there was no urgency felt concerning those areas which were primarily agricultural. Also, whereas Britain had had a tradition of local government reform, Canada had not, and it is possible that Ontario was simply feeling its way.

Each of the commissions had to be given areal terms of reference and these were always in terms of the already existing subdivisions. In the Hamilton-Burlington-Wentworth Review Commission with which I was involved officially, Hamilton was the central city, Wentworth the surrounding county and Burlington a large suburb on the opposite shore of Hamilton Harbour but in another county. Most of the commissions were named after pre-existing counties. This nomenclature had the effect of fitting the new units into the boundary constraints of the old system.

Over 120 delegations submitted briefs and appeared before the Steele Commission, as the Hamilton-Burlington-Wentworth Review Commission came to be known. The planners and the directors of the larger services, thinking in functional-structural terms, were in favour of the unification of the area, with as much power as possible allotted to the top tier, the Regional Council. The City of Hamilton, which included two-thirds of the total population, supported a one-tier system, or the unification of the area under one council, the membership of which would be based upon representation by population (City of Hamilton 1968).

The municipalities outside the city limits presented a totally different interpretation. Fundamentally, they saw no reason for change. The old system was working; they were getting along well without the city, cooperating where necessary. They recognised that since they were within the county, their municipalities were certain to be included within the new Region, but they exhibited an almost paranoic fear of the central city. They urged that there be no extension of the city limits, and that representation on the future Regional Council be so weighted that the city not have a majority of the members (Yates and Yates 1969). They were in the curious position of invoking democracy, specifically self-determination, to assure their continued existence, and of objecting to democracy, specifically representation by population, in the administration of the Region. They assumed that all the councillors from the city would vote as a bloc against the suburban members on important issues.

They possessed a strong sense of local self-identity. By European standards these were young units but by Canadian standards they were time-tested and traditional; each seemed to be able to call up memories of brave pioneers who had fled from the American Revolution or from starvation to create a new British colony in the North. One must conclude that history is relative; one must place duration within the perceived span of time. Areas which can be traced back to their beginnings are often imbued with a passionate sense of territoriality, whether the origins were in 1000 B.C. or 1800 A.D.

When it became clear that the amalgamation of the rural areas was inevitable, their inhabitants retreated to that final link with their territorial past, the local name. The sacred name should not be deleted from the map. The result in Ontario was a number of hyphenated double names: Hamilton-Wentworth, Ottawa-Carleton, Haldemand-Norfolk.

The farmers in Hamilton-Wentworth asked for special consideration. Although they occupied some two-thirds of the land area, they comprised less than five per cent of the population. They feared a lack of understanding on the part of the urban-based councillors and planners; they feared increased taxes and restrictions. It is fair to say that they distrusted planners instinctively. They feared that the urban bias of their clientele would lead the planners to look upon the rural areas not so much as productive farmland with its own concerns and problems, but as picturesque countryside, as open space or green belt, and as potential recreation land to be kept free of annoying smells, unsightly structures and pesticides. They feared that urban evaluations of their land could remove their right to sell their land at advantageous prices and, carried to an extreme, make them virtual serfs upon the land.

The principal suburban municipality, Burlington, lay outside the limits of Wentworth County. Its delegation maintained that since it was outside the county, it was outside the area to be considered. In addition, it was argued that the continuing viability of its own county depended on Burlington's inclusion within that county (Town of Burlington 1968). On the basis of the close economic and social interconnections between Hamilton and Burlington, the Steele Commission recommended the inclusion of Burlington in the proposed Region, a decision that had been strongly advocated by the smaller municipalities who wished to have a counterweight to the dominance of the central city (Steele 1969). In defiance of the government's refusal to allow referenda on local government, the politicians of Burlington called a snap election on the issue and the vote was ten to one against entry into the Hamilton Region (Burlington Post, 3 and 10 December 1969). Using this 'democratic' result and intense political pressure, Burlington was successful and had its county transformed into a Region.

The question which would seem to underlie all decisions as to limits, namely, 'What is the optimal size of a local administrative unit?', was not given much serious consideration in Ontario. Echoing statements of various British reports, figures of over a quarter of a million were quoted for a Region, and from 10,000 upwards for a subdivision, but the variations in population distribution made the adherence to any norms difficult. In practice the new Regions were established <u>a priori</u> within the framework of the old county system, and with but rare and minor exceptions, the county boundaries became the new Regional boundaries. This was true even in the vicinity of Toronto where the functional city region came to be divided among four or five administrative Regions. (It is also true that placing all of Toronto city region within one Region would have created a unit of extraordinary complexity and political power.) In effect, Regional Government in Ontario is a reformed version of County Government, with the important change that the city and county are no longer separate.

Opposition to the introduction of the new system was strong and there can be little doubt that outside the cities themselves, the imposition of the new system was unpopular. Many of the rural Conservative Party supporters felt betrayed by the Government that they had helped to elect. In the two elections which followed, (18 September 1975, 9 June 1977), the Conservative Party lost its majority for the first time in over 30 years. Although the Conservatives have now regained their customary majority position, they achieved this on the basis of overwhelming support from Toronto (19 March 1981). In a curious way, what had been preeminently a rural-based party has become a party based on the capital metropolis. As a result of this tangible opposition, the provincial government stopped all attempts at Regionalisation in 1974. Thus most of the land in Southern Ontario is still under the county-township system, but a majority of the people are now living in the Regions.

Within Hamilton-Wentworth the enforced union of the city and county led to an increase in hostility. The rural- suburban municipalities were continually wary of any attempts by the central city to exercise its dominance or to thrust its problems upon the Region. After five years of unhappy operation, the provincial government established another commission, the Stewart Commission, to recommend how this Region could be brought to work smoothly. The Stewart Commission recommended the eradication of all the internal subdivisions and their merging into a single city, coterminous with the former county (Stewart 1978). The resultant uproar was a magnificent display of local democracy at work, causing the provincial government to shelve the Stewart recommendations indefinitely.

The history of these events points out the position of Canada between the United Kingdom and the United States in terms of political theory and practice. Canadians are a people motivated by American perceptions and expectations but operating within a British political structure. As in Britain, the higher level of government, in this case the province, has the power to alter local units at will, but as in the United States, the citizenry believes strongly in local democracy. Adjustments to administrative areas can be made when and where they are perceived to be needed, but the government must always reckon with the opinions of an articulate local electorate.

Although I have dealt principally with Hamilton- Wentworth, there are other Regions of varying types in Southern Ontario, and these should be looked at briefly.

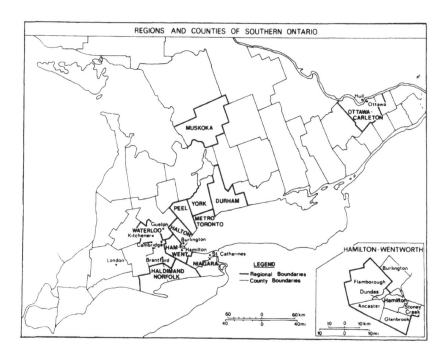

Figure 17.1: Regions and Counties of Southern Ontario

There are two examples of what approximate to the classical model of the nodal or city region - the central city, its ring of suburbs and the umland. In both cases, however, the functional city region extends far beyond the administrative boundary. These two are Hamilton-Wentworth and Ottawa-Carleton. Curiously, in both cases the largest suburb has been omitted. The case of Burlington has already been noted; Hull, the French alter-ego of Ottawa, lies in Quebec. Both Regions have seen conflict between the central city and the other municipalities.

Most of the Regions include two or more cities or large towns. In all cases, competition rather than cooperation has been the rule. The smaller places are vigilant lest the largest place obtain the primary facilities; the largest place is concerned as it pays more in taxes than it receives in benefits or power. Niagara, which includes five cities, has had an especially turbulent history. St. Catherines in Niagara, and Cambridge in Waterloo have both made attempts to secede from their Regions, while other places such as Burlington in Halton and Flamborough and Dundas in Hamilton-Wentworth have expressed their desires to secede (Spectator, 1971, 1976, 1980,1981).

One Region to the south of Hamilton, Haldemand-Norfolk, is a union of two agricultural counties. Both names had to be maintained and except for the fact that they are both within the hinterland of Hamilton, they have little in common. Upon unification, they could not agree on a location for the new administrative centre, hence the Regional apparatus was divided in two and located in the two former county seats, which are 40 kilometres apart.

An overall criticism which is heard everywhere is that the introduction of Regional government has not led to greater economies but greater costs. The expected savings from the removal of duplications were more than offset by the creation of new structures which demanded massive new facilities employing numerous high-salaried professionals. Bigger may be better and more modern but it is not cheaper. At first, the provincial government maintained that the upsurge in costs was transitional but it later retreated to the explanation that a twentieth century urban economy cannot be managed with a 'horse and buggy' governmental system.

IMPLICATIONS

Many geographers and planners tend to see the world in terms of foci and networks of interconnections that is, in terms of the nodal region, or Scottish model. The terms city region, growth pole, core and periphery have become commonplace. The urban environments within which we live and work give us, almost automatically, a centralist outlook. Instinctively, we look outward from our central places and judge the hinterland in terms of its relationship to the central place, forgetting that few want to be merely a part of a web centred on some external point. The umland does not think of itself as umland, but rather as a collection of vital centres, each saturated with history and personality.

Because we think in terms of connectivity, we tend to underestimate the continuing strength of the rural-urban dichotomy. Despite the revolutions in transportation, the person 'living in the country' often thinks of himself as rural, even if all his economic and social ties are with a nearby urban place. The city walls are still in the mind, if not on the ground.

Although unfortunate, the intense fear of the central city must be faced by anyone proposing reforms in local government. The enforced joining of the city with its immediate surroundings has tended to exacerbate fear and animosity. The fear of the loss of local identity appears to be stronger and more widespread than realised. It is certainly rife in Canada, the United Kingdom and the United States.

Our models have been based on the measurable, the quantifiable, the readily available, the obviously visible. The more subtle aspects of reality have often been overlooked or, if recognised, deemed to be less relevant to the problems of planning than the 'hard data'. The primacy of efficiency has been accepted as self-evident. As Rees (1971) has commented, in the British context,

> 'the commissions were, from the beginning, inclined to pay little attention to democracy as opposed to efficiency, to the convenience of representatives as opposed to the convenience of the specialists, . . . [the Commissions] always appeared to be searching for a minimum size for efficiency rather than a maximum size for democracy.'

It may be that the time has come for political geography to return to three sets of perceptions which were once central to the discipline but which were jettisoned during the 1960s: the distinctiveness of political units, the importance of the precise location of boundaries and the significance of local loyalties.

Maps drawn before the Second World War often emphasised the subdivision of space into political units, almost to the exclusion of other details and distributions. Depending on scale, every state or province, every county, or every municipal area was given its own colour which differentiated it from its neighbours.

This sense of the discreteness of small-scale political areas was one of the earliest casualties of the ceteris paribus principle which seemed to underlie early attempts at model-building in geography.

The study of boundaries has also suffered a notable decline since the 1950s. The stabilisation of international boundaries by Article 2 of the United Nations Charter seemed to undercut any need for further study of the delimitation of boundaries. As a consequence, boundaries are commonly examined now in terms of their effects on flow patterns or on the exercise of specified functions, but rarely in terms of their placements (Short 1982; Brunn 1974; Minghi 1963). In local government reform, the location of the outer boundary of a region is rarely treated as a matter of great concern to planners; rather it is perceived as the unavoidable external horizon of future activities. Arguments may flare as to the areas to be included or excluded, but rarely as to where in detail the limits are to be drawn. Internal boundaries are seen as jurisdictional inconveniences to be superseded where possible.

To the local inhabitants, however, the locations of these limits are matters of concern. The boundaries form the frames for their areal consciousness and of their territorial identities (Ullman 1938). In a psychological sense, the boundary represents that perimeter which marks off the rest of the world. The residents of a municipal unit will tend to feel an inward boundary adjustment to be an amputation, an amalgamation with another unit to be a change from concreteness to amorphousness and an integration into a more powerful unit as a change from rootedness to deracination. This does not mean that boundaries should never be changed; it does mean that the same attention to detail should be given to the delimitation of boundaries at the local level. The placement of boundaries should not be viewed as merely a by-product of the identification of functional regions.

Until recently, local patriotism was deemed a laudable pursuit (Freeman 1968: 17-18). Loyalty to a local area , such as a New England township or a Swiss canton, was understood to form a basis for loyalty to a nation. However, attachments to small-scale local areas have so often hindered the efforts of planners that these ties have tended to be looked at with disfavour. Such inward-looking loyalties have been viewed as archaic and a hindrance to administrative progress. It is here that the battle between 'democracy', taken to mean smallness, and 'efficiency', taken to mean bigness, has often been waged (Johnston 1979: 141-58; Scottish Grand Committee 1972: 255).

At times the charge is heard that a true local loyalty scarcely exists in our mobile society. Calls to tradition are held to be no more than vocal displays by politicians and vested interests who wish to hold on to their petty fiefdoms. Experience does not support this charge and local politicians have sometimes shown a reluctance to do battle with the higher, more powerful levels of government. The most strident opposition to local change has come from ad hoc groups, organised and composed of non-elected citizens (Valley Journal, Dundas Star, both May and June 1978).

Many in the Western World have centred their areal and political loyalties on two levels, the nation and the locality. It is significant that generally there is little attachment to medium-scaled units. A major problem facing the new Regional governments of Ontario is the absence of any appreciable loyalty to the Regions (Spectator, 28 May 1981). Residents and politicians continue to focus their loyalty upon their local subunits. Perhaps with time, a sense of attachment to the new, larger 'local' units will develop. However, it bears noting that whereas the upper level loyalty has expanded through the centuries from clan or barony to the nation-state, the lower level has remained fixed on the commune, village, town, city. Certainly any new administrative structures which are to be devised should make allowance for this continued human attachment to a set of territorial cells.

Having said these things, it must nevertheless be repeated that the old system did require reform. The shortcomings summarised at the beginning of this chapter have cried out for rectification. Some method of achieving an approximate equalisation of taxes, responsibilities and privileges within metropolitan areas, should be sought. A kind of overall planning of the metropolitan area seems essential.

The two-tier system, which is used in Ontario, appears capable of serving these needs well. However, to allow for the maximum local democratic participation, the lower tier units should remain truly local; they should not be amalgamated into extensive areas in order to meet some arbitrary threshold figure (Scotland, Written Evidence 1967: 24). If 'efficiency' is to be the dominant criterion used to define the upper tier, then the lower tier at least should be left to the full exercise of local 'democracy'. It has even been argued that the preexisting cities should be subdivided into more manageable units (Wheatley 1969: 189; Scottish Grand Committee 1972: 54, 62). The identification of suitable lower-level units and the locations of the boundaries between them call for considerable research.

The method of allocation of powers between the two tiers should be reversed, with as much responsibility as possible left to the lower tier. At present, the tendency is clearly to place all major authority in the upper tier or beyond, and to allow the lower tier only some peripheral participation in the operation of the Regions. Government from the centre will usually be distrusted, no matter how good its intentions might be. Local participation in decision-making has both practical and idealistic benefits. A combination of the best features of the traditionalist (American) and homogeneous region (English) models would appear to offer the best prospects for success.

One final, practical point should be made. If something as radical as a reorganisation of local government is to be effected, it should include the entire political area. At the present time in Southern Ontario, Hamilton and Kitchener are 'Regionalised', but neighbouring Brantford and Guelph are not, nor are the large cities of Windsor and London. Haldemand-Norfolk is Regionalised but some neighbouring agricultural counties are not. One can scarcely blame people for looking across the Regional limits to the picture of an idealised past just outside. Simultaneous reorganisation would allow for the delimitation of optimal boundaries.

It seems clear that in planning and implementing our new administrative structures, we cannot think only of economic or managerial efficiency and operate in terms of abstract models. Deeply rooted local feelings must be taken into account; local territorial self-identities should not be dismissed as anachronistic emotionalisms. Unfortunately, the local citizenry has often tended to be alienated by an overemphasis on purely rational solutions.

REFERENCES

BLAKE, W. [1965]
    'And did those feet in ancient times', in D.V. Erdman
    (ed.), The Poetry and Prose of William Blake, (Doubleday,
    New York), 95.
BRUNN, S.D. [1974]
    Geography and Politics in America, (Harper and Row, New
    York).
BURLINGTON POST
    10 December 1969, (Burlington).
CITY OF HAMILTON [1968]
    Submission [to the Hamilton-Burlington-Wentworth local
    government review commission], (City of Hamilton).
DICKINSON, R.E. [1964]
    City and Region: A Geographical Interpretation, (Routledge
    and Kegan Paul, London).
DICKINSON, R.E. [1967]
    The City Region in Western Europe, (Routledge and Kegan
    Paul, London).
DICKINSON, R.E. [1970]
    Regional Ecology: The Study of Man's Environment, (Wiley,
    New York).
DUNDAS STAR
    May-June 1978 (Dundas)
FREEMAN, T.W. [1968]
    Geography and Regional Administration, (Hutchinson,
    London).
JAMES, J.R., J.W. HOUSE and P. HALL [1970]
    'Local government reform in England: A Symposium',
    Geographical Journal, 136, 1-23.
JOHNSTON, R.J. [1979]
    Political, Electoral and Spatial Systems, (Clarendon Press,
    Oxford).
MINGHI, J.V. [1963]
    'Boundary studies in political geography', Annals of the
    Association of American Geographers, 53, 407-28.
REDCLIFFE-MAUD, THE RT. HON. LORD [1969]
    Royal Commission on Local Government in England 1966- 1969,
    Report, (CMND 4040,HMSO, London).
REES, I.B. [1971]
    Government by Community, (C. Knight, London).
ROBSON, W.A. [1968]
    Local Government in Crisis, (George Allen and Unwin,
    London).
ROYAL COMMISSION ON LOCAL GOVERNMENT IN SCOTLAND [1967]
    Written Evidence, Minutes of Evidence, (HMSO,
    Edinburgh).

SCOTTISH GRAND COMMITTEE, PARLIAMENTARY DEBATES, COMMONS, OFFICIAL
REPORT
    Local Government (Scotland) Bill, Session 1972-73,
    (London).
SENIOR, D. [1965]
    'The City Region as an administrative unit', Political
    Quarterly, 36, 82-91.
SENIOR, D. (ed.) [1966]
    The Regional City, An Anglo-American Discussion of
    Metropolitan Planning, (London).
SENIOR, D. (ed.) [1969]
    'Memorandum of dissent', in Redcliffe-Maud, Royal Commission,
    Volume II.
SHORT, J.R. [1982]
    An Introduction to Political Geography, (London).
SMITH, L.J. [1967]
    Report (Ontario Commission on Taxation, Toronto).
THE SPECTATOR
    1-9 December 1969, 21 January 1976, 12 and 27 May 1978,
    3,15,16,20,21 and 23 October 1980, 25-28 May 1981, 22
    September 1981, 3 November 1981, (Hamilton).
STEELE, D.R. [1969]
    Report and Recommendations by D.R. Steele (Hamilton-
    Burlington-Wentworth Local Government Review) (Ontario
    Department of Municipal Affairs, Toronto).
STEWART, H.E. [1978]
    Hamilton-Wentworth Review Commission Report, (Government
    of Ontario, Toronto).
TOWN OF BURLINGTON [1968]
    Submission [to the Hamilton-Burlington-Wentworth local
    government review commission], (Burlington).
ULLMAN, E.L. [1938]
    'Political geography of the Pacific Northwest', Scottish
    Geographical Magazine, 54, 236-39.
VALLEY JOURNAL
    April-June 1978 (Dundas).
WHEATLEY, THE RT. HON. LORD [1969]
    Royal Commission on Local Government in Scotland 1966-
    1969, Report, (CMND 4150, HMSO, Edinburgh).
YATES E. and G. YATES [1969]
    Representations on behalf of the County of Wentworth, [to
    the Hamilton- Hamilton-Burlington-Wentworth Local Government
    Review Commission], (Hamilton).

Chapter 18
VERSATILITY VERSUS CONTINUITY — THE DILEMMA OF JURISDICTIONAL CHANGE

Rex D. Honey

INTRODUCTION

Most countries have dealt in some way with the perceived
inadequacies of their systems of local and regional governance
during the years since World War II. Some have sought to match
the emergent metropolitan reality of economic and social life
with a metropolitan scale of government. Others have tried to
add a regional scale of government to a federal structure. All
of these countries have experienced battles over the redistribution
of territory and responsibilities among governmental authorities.
Often, the conflict has been concerned with what authorities
there ought to be. Typically, the number of local authorities
has fallen, sometimes sharply. The number of school districts
in the U.S.A., for example, shrank from over 120,000 in 1930 to
below 20,000 by 1970 (Sher and Tompkins 1977: 44).

At the heart of these jurisdictional changes have been
geographical issues. As the British political scientist R.A.W.
Rhodes so aptly described the situation, 'The key question to
ask about local government reform is: How has the reform altered
the existing areal distribution of power?' (Rhodes 1979: 576).

The purpose here is to answer Rhodes' question. First, the
nature of the jurisdictional organisation of space is examined,
followed by consideration of two major goals of combatants:
versatility so that governments are flexible enough to respond
to problems which may arise and continuity so that services are
not disrupted and identities not sacrificed. The paper concludes
with brief examinations of local government reorganisation in
Wales and school district consolidation in the U.S.A.

The nature of jurisdictional space

Unlike other types of spatial organisation, jurisdictional
space is necessarily compartmentalised. The spatial relations
of a residence will illustrate this point. A residence might be
in the service areas of several shopping centres or the commuting

field of two or more employment centres; however, this same residence is within one and only one country, as well as one and only one unit at each level within that country's jurisdictional hierarchy. In the U.S.A., this would be one state, one county, one school district, possibly one municipality. Jurisdictional space is thus Platonic space in that it is bound and absolute (Lukerman 1961: 201) and is thus all-encompassing, with no place being left out; it is container space.

Economic and social space are, in contrast, unbound, overlapping and fluid. Rather than being defined by boundaries they are defined by the relationships of points within space - a firm to its customers, friends to each other. They are Aristotelian rather than Platonic, relative rather than absolute. Moreover, they are dynamic, changing each time firms change their marketing strategies or individuals their friendship patterns. These changes are frequent, virtually continuous and different for everyone. Jurisdictional space, on the other hand, changes infrequently and only with great difficulty; and, of course, jurisdictional space is the same for everyone, finite.

The nature of the jurisdictional organisation of space is derived from the nature of the state itself. The territorial-society definition of a state is, in reality, more a continuum than a dichotomy. Anyone born in the U.S.A. is automatically an American citizen; those born in West Germany to Turkish parents are not German, however. The specific nature of jurisdictional borders depends on the nature of the states involved.

Similarly, the functions of subnational governmental authorities determine the spatial significance and characteristics of those authorities. Are divisions simply field service areas for the performance of centrally controlled services or are they actually units of self-government with internally generated funds and determined policies (Fesler 1949)? Requirements of the two differ markedly given that differentiation of both demand and provision is only necessary with self-government (see Dear 1981; Clark 1981; Massam 1975; Lea 1979).

Perceptions and expectations from local government vary over time and space. As societies change, pressures for institutional change emerge, among them pressures for territorial change. These pressures are opposed by groups which benefit from the status quo and risk losing power if change occurs. We shall consider reasons for change as well as those for keeping things the same.

VERSATILITY: MATCHING PUBLIC SPACE TO PRIVATE ACTIONS

'By common consent, local government boundaries should be changed every ten years or so.' (Freeman 1968: 22). Not everyone agrees that local government boundaries need ever be changed, let alone every ten years. Nevertheless, Freeman makes a very important point: local government areas are established under one set of geographical conditions and tend to remain in force long after those geographical conditions have changed. The rationale for a particular set of boundaries and distribution of responsibilities may have vanished with expanding urban growth or improvements in transportation, but the boundaries and distribution of responsibilities may survive intact even though rendered quite inappropriate by the changed circumstances. Put another way, the jurisdictional organisation of space may be appropriately designed to reflect economic and social geography but as these change, the jurisdictional organisation becomes less appropriate. Needed is a governmental system suited not only to the economic and social conditions at the time of the system's creation but which also has the versatility to adjust as those conditions alter.

Rowat (1980: 594) cites societal change as the main reason for jurisdictional change. In particular, he cites urbanisation and industrialisation as the reasons for which old government systems have not worked well. As cities expanded becoming metropolitan complexes, the old institutions failed to cope. Consequently, many states created new metropolitan institutions. Likewise, industrialisation brought about a change in the relationship between society and polity. People came to expect and receive new services and roles from the public sector. With government thus playing a larger role, the need for government to work well became more compelling.

Changes in the public sector in the twentieth century have been multi-faceted. All societies, from those purporting to eliminate the private sector to those professing to have saved private enterprise, have been affected. Whereas at the turn of the century most transportation, electricity, health care and education services were provided commercially, the public providers now serve the majority of the beneficiaries of these services.

Even where governments do not provide services directly, their influence is felt through regulations. Citizens of many societies expect greater government protection and assistance than ever before - against flooding and other natural hazards; against crime and other anti-social behaviour and against unemployment and other catastrophes.

Consequently, taking all these factors together, most societies have called upon their public institutions to do more and do it better. It is little surprise to find that the institutions often fall short of public desires. New societies, especially new urban societies and their leaders, have perceived their institutions, including those concerned with the territorial structure of public authorities to be inadequate. L.J. Sharpe (1979: 18) has chided social scientists for understating 'the effect of space on the distribution of power', leading to an examination of space and power.

Geographical problems

In the vast literature on the problems of local govermment, a common theme is that local governments are too small to carry out their assigned functions - too small to be efficient, equitable or effective.

Analytical arguments that governmental units are too small tend to follow two main thrusts. One is that authorities cannot attain economies of scale because they serve too few people and have too few resources. For particular services, jurisdictions must attain a minimum population threshold if benefits are to surpass costs or in order to extract maximum output for a given input. Hence Conant (1959: 80) argued that a minimum of 100 children per grade was required for the operation of a comprehensive high school and the Herbert Commission, considering reorganisation of local government in Greater London, argued for a threshold of 100,000 for lower tier authorities in a two-tier metropolitan structure (Royal Commission on Local Government in Greater London 1960: 232).

The second thrust is quite different. Its argument is that a governmental unit must have jurisdiction over the necessary area if it is to do a specific job effectively. Thus an authority with responsibility for traffic management should control the territory needed for managing traffic, i.e., the area of significant commuting. Similarly, an authority with responsibility for flood control might be given jurisdiction over a drainage basin. This argument calls for 'internalising externalities', effects of government action which fall beyond the boundaries of that government.

However, proponents of jurisdictional reorganisation do not just argue that jurisdictions are too small; they often argue that jurisdictions are not small enough. These arguments also tend to divide along two main lines. The first has to do with accessibility. Simply put, it is that local government be local, i.e. close to the people. Residents of the Shetland and Orkney Islands, for example, argued successfully for their own region rather than inclusion in a larger region when Scotland's reorganisation was being planned (Reform of Local Government in Scotland 1971: 14).

A quite different argument in favour of small authorities is based on differences that exist within a population in demand for public goods and services, as well as differences in the style of government desired. In effect, these arguments ask that different sectors of the public be allowed to make different choices. If individuals with different preferences are territorially segregated, then they can be divided into separate jurisdictions maximising the correspondence between preference and level of services actually provided (Archer 1981).

Despite the tendency for local government boundaries or, indeed, an entire jurisdictional structure, to become less appropriate over time, powerful forces against jurisdictional change generally protect the status quo. Amongst these are historical considerations, the disruptive effects of jurisdictional change and the opposition of vested interests such as those who might lose power. Each of these factors will be addressed separately.

Soja (1971: 34) has argued that a sense of spatial identity is a major dimension of societal territoriality. While he was referring primarily to national identity, people also gain a strong sense of identity with jurisdictions ranging down to local levels. The merger of school districts in the U.S.A., for example, has been impeded by the allegiance people have to their communities. If merger is perceived to threaten community identity, it is opposed.

Traditional jurisdictions proffer a sense of identity as well as the comfort of continuity by linking the present with the past. For many people, the old units carry a view of who we are and how we have come to be.

Much of the opposition to England's recent local government reform stemmed from desires to maintain local identities. The reorganisation threatened identities with origins in Norman times. When facing the prospect of losing their identity, residents of Herefordshire marched a steer around Westminster. The marchers did not save the exclusiveness of their county but they did succeed in maintaining a modicum of identity, for the final name of their new county is 'Hereford and Worcester' rather than the 'Malvernshire' which had been proposed.

Overhauling a local government system necessarily creates problems of transition. Power must be transferred from one elected body to another and one set of appointed officials to another. Perhaps more importantly, a uniform level of services will have to be provided where perhaps several existed in the past. Again, calling upon the English experience, the new level of services is likely to be at the highest level of service among the authorities prior to reorganisation. These higher services bring with them, though, higher costs which jeopardise economies which the reforms aimed at achieving. Also, the new identities require a reorientation of the population with individuals having to learn a new set of institutions which must be used for solving problems (Norton 1979: 273).

This opposition is likely to be strongest amongst those with the most to lose. Recalling Rhodes' comments about redistribution of power, one expects and observes opposition from those who have power but stand to lose it. Hence, the Labour Party opposed the creation of the Greater London Council in the early 1960s because Labour stood to lose its hegemony over London government (Smallwood 1965: 189-93). Similarly, many officials in suburban St. Louis opposed the referendum on metropolitan government there (Schmandt et al. 1961: 37).

The inertia of the vested interests of the status quo is a powerful brake on jurisdictional change. The strength of this depends largely on the way decisions are made. The more groups which must approve change, the less likely change is to occur, because the opportunities for successful opposition are greater. Consequently, the U.K., with its centralised parliamentary system, has reorganised local government across Great Britain. The U.S.A., at the other extreme, has achieved only modest change, due largely to the decentralised manner in which it deals with local government (Honey 1981).

POWER SHIFTS IN JURISDICTIONAL CHANGE

A new local government system commenced operation in Wales in 1974. Although it did not represent a complete break with the past, the new system incorporated significant territorial and functional adjustments. The Welsh experience illustrates well the tension between the jurisdictional organisation of space, which is rigid by nature, and effective management of a complex society, which requires flexibility.

Old Jurisdictional System

The Welsh system of local government followed the lines of the English system and the new Welsh structure is consistent with that in England. The main authorities were the counties, each of which was subdivided into several districts, some of which were further subdivided into parishes. Some of the counties dates from the thirteenth century, others from the sixteenth, when Wales united with England. Administration of the counties changed sharply, but territorially the major change was the designation of county boroughs which were to operate independently of the county structure. Thus the Welsh cities - Cardiff, Swansea, Newport and later, Merthyr Tydfil - separated from the Welsh countryside. By the 1960s, Welsh local government comprised thirteen counties, four county boroughs and more than 180 districts, all to serve a population of some 2.7 millions (Figure 18.1).

Geographical problems

Scholars and government officials cited specific flaws in the local government system. These flaws have to do mainly with what Sharpe termed socio-geography and service efficiency (1979: 35). In other words, the jurisdictional structure of Welsh authorities made operation of local government needlessly difficult because the spatial structure bore insufficient relationship to patterns of social and economic interaction, and because too many authorities were too small, especially in resource base, to execute their responsibilities with anything approaching efficiency. In other words, the Welsh authorities failed to internalise externalities and they failed to reach service thresholds.

The problems of socio-geography had two elements. First, the influence of Welsh cities, like all modern cities, had expanded markedly, both as a result of actual urban growth and of improved transportation and communications. So, areas previously distant from Cardiff found themselves within commuting range. Areas beyond Swansea's service area fell within that city's sphere of influence. Managing a contemporary metropolis was difficult as long as the central city was divided from its suburbs and hinterland.

The second factor of socio-geography was tied to the mountainous nature of Wales. As the rail and road systems evolved, spatial relations changed. Requirements of rapid communication ran against the grain of the old boundaries in a number of instances. Reorientation was required so that the jurisdictional organisation of space could better match the economic organisation of space.

Reform Process

The chain of events eventually culminating in a new local government system for Wales began with the appointment of a Local Government Commission for Wales in 1959. With an objective of securing 'effective and convenient local government', the Commission considered spatial and operational changes (Local Government Commission for Wales 1962: 3). Limited by its warrant, the Commission recommended a structure with seven counties in 1962, but had to leave the county boroughs alone except for some modest boundary extensions.

The process continued in 1965 when the Secretary of State appointed an Interdepartmental Working Party which was given a freer hand in considering fundamental change. Deliberations continued until reorganisation came to fruition with passage of the Local Government Bill in 1972. This legislation, which transformed local government in England and Wales, not only restructured local government boundaries and duties, but recodified local government law and administrative arrangements.

Solving the Problems

Strengthening local government so that decisions could be made by local rather than national officials was one of the major objectives of the reform. The twin organising principles of the

234

Figure 18.1: Wales – Before Local Government Reform (left) and After (right)

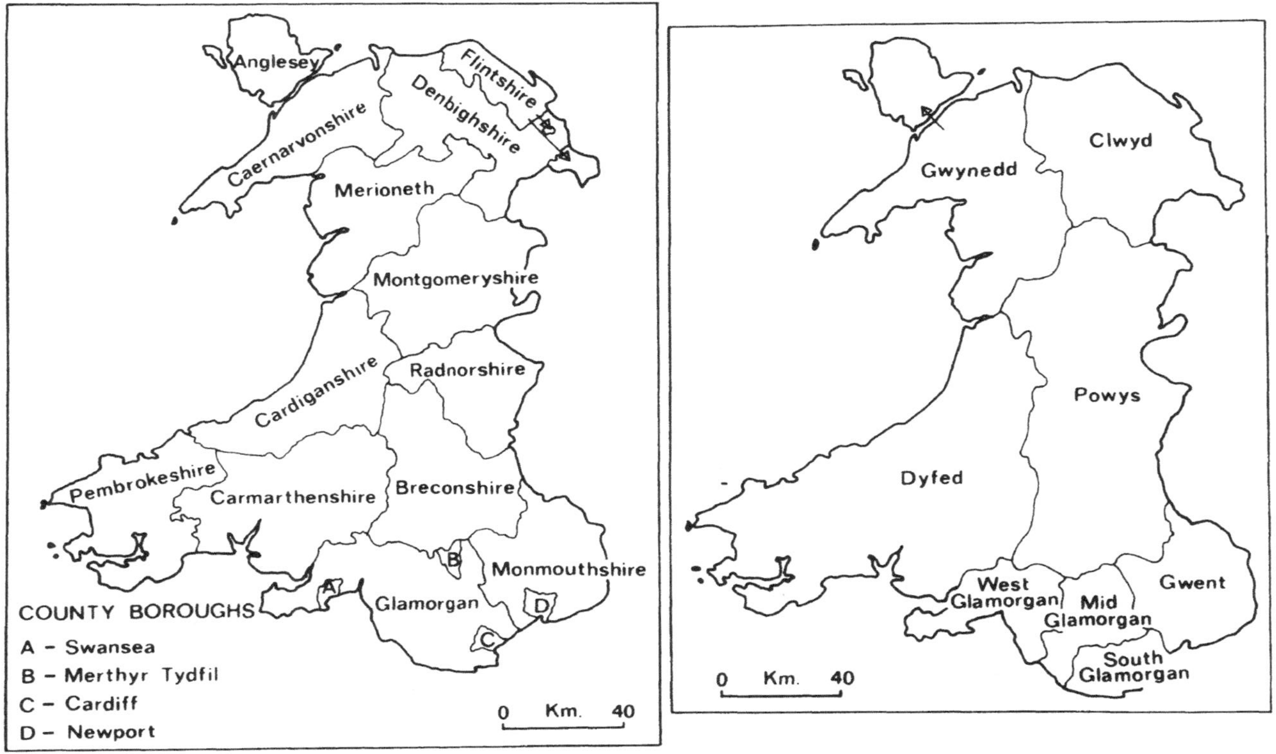

235

new system were: a threshold of 200,000 people so that the new counties could operate effectively and efficiently; and joining areas which need to be managed together, or internalising externalities.

The new structure has eight counties and 37 districts, less than one-quarter the number of local authorities in Wales before reorganisation (Figure 18.1). Community Councils do serve smaller areas but they are consultative rather than decision-making bodies. County boroughs have been eliminated. Unquestionably, overseeing local government has been made easier for ministers and bureaucrats in Cardiff and London. Ministers and civil servants encountered several difficult problems in designing the new jurisdictional divisions, however, and several of these problems merit attention.

In South East Wales the major dilemma was in deciding how to join the county boroughs and counties. The population of Wales is concentrated in the south, more specifically in the southeast. Consequently, the Local Government Bill of 1972 provided for four counties in South East Wales, based on Newport, Swansea, Cardiff, and a fourth consisting largely of the mining valleys leading to Cardiff. Each easily met the 200,000 threshold, but the principle of joining places tied together by daily interaction was sacrificed to let people with different interests have their own county. In effect, the minister preferred a public choice argument to internalising externalities.

The rest of Wales - 40 per cent of the people, 80 per cent of the area - now consists of four counties. One, Powys, required a major relaxation of the threshold requirement, with a population of only 100,000, even though it is the second largest of the Welsh counties in area. The terrain is mountainous, transportation difficult. In all, three counties were merged to form Powys. But, despite the population problem, the southern fringe was assigned to Mid-Glamorgan because the area involved has stronger ties to its south. Accessibility won over the threshold requirement in this case.

The South West presented different problems. In addition to a transportation system limited by mountains, this section had strong cultural divisions. Pembrokeshire was known as the English Vale because of its long dominance by an English population. The people of Pembrokeshire, with almost as many people as Powys, wanted to retain their own county. The minister felt that the benefits of a more powerful Dyfed (the new county) outweighed those of a resource-poor Pembrokeshire. In this instance, the threshold principle reigned over the public choice argument (Thomas 1975).

In the north of Wales, two counties serve where five existed before. To the west, Gwynedd, only slightly surpasses the threshold, and then only because boundary adjustments gave it some of the territory from the two old counties which form Clwyd, the new county to the east. The boundary adjustment brought together everyone in the Conway Valley, which could have been

236

assigned to either of the new counties. The threshold need of Gwynedd swung the assignment in that direction. In this case, the absolute nature of jurisdictional space meant that one authority had to lose territory because another needed population.

Redistribution of Power

While the Welsh local government reform is a centralisation of local government, it should also produce a decentralisation of power. Local government units are now much larger and have much greater tax bases. The new authorities are not sufficiently large to carry out responsibilities which used to be impossible for them, either because they lacked resources or did not control the requisite territory. Welsh local authorities, for example, control sufficient territory to permit effective planning for the future, with the possible exception of the area north of Cardiff.

To a large degree, rural independence is now subject to urban influence. The larger city votes can outweigh rural votes, at least in Gwent, South Glamorgan and West Glamorgan. Those whose paths cross daily are generally sharing political decisions.

The Welsh reform certainly carried the imprint of history. The new county names hark back to ancient Wales. The Welsh reform was an organised exercise, accomplished finally with a single, sweeping action. The impact of a decision in one area was considered in light of the consequences in other areas.

SCHOOL DISTRICT REORGANISATION

Without any question, school districts are the jurisdictions which have changed the most in recent decades in the U.S.A. Some 127,000 independent school districts existed in 1932, as compared with fewer than 16,000 in 1982, a reduction of 87 per cent. Yet all this occurred without any direct action by the federal government.

In the American governmental system education is a state preserve. The federal government provides funds for a large variety of educational programmes and federal courts require school districts to work in accordance with the American Constitution, but the states have constitutional responsibility over education. Rather than provide schools directly, though, the states delegate that responsibility to school districts. Sometimes, especially in the East, municipalities run the public schools, but the American norm is for school districts to be special authorities concerned only with education.

Universally available public education became standard practice in the U.S.A. in the nineteenth century. Free public education is in fact fundamental to the American sense of fair play and opportunity. Early schools tended more to be authorised, rather than supported, by the states. Traditionally American public schools have obtained their financing from local property taxes. Given the juxtaposition of valuable property and children of school age, the borders of the school districts have been very important. Frequently, school taxes surpass municipal and county levies against property, so school district boundaries have been highly politicised, district residents often seeking to increase property value while inhibiting the addition of children. Only in recent years, as a result of court decisions and state legislation, has school funding been separated from local property so that such fiscal squabbling can be averted.

Even after continual rounds of school district reorganisation, most school districts still serve basically rural or small town populations. Before the major waves of consolidation the vast majority of districts were rural. It was typical of nineteenth century America that relatively small rural areas, perhaps a few square miles, would have their own schools.

Reorganisation Ethic

Even as the Plains and West were being settled, pressures for changing the spatial structure of education arose and a reorganisation ethic developed as a response to fundamental societal changes. The urbanisation and industrialisation which Rowat credits for much of the world's local government reform applies also to American school districts. Urbanisation and the mechanisation of agriculture combined to generate an exodus from American farm land. Within a generation of being settled many areas in rural America began to lose population. Almost all rural areas experienced net outmigration, a pattern which has only been reversed in the last decade and even then, generally near metropolitan areas or amenity sites.

As late as 1930 the U.S.A. had about 150,000 one-teacher schools. The number had dropped to 15,000 by 1961 and is now a fraction of that (American Association of School Administrators 1962: 10). Clearly, the argument went, rural children must come together in larger schools or face severe handicaps in the complicated present and even more complex future.

Perhaps the biggest boost for maintaining the school consolidation momentum came from Conant (1959). He called for comprehensive high schools so that all children could develop to their potential. Arguing for the universal availability of vocational, general and college preparatory curricula, Conant concluded that to be both equitable and efficient high schools must have at least 100 pupils in each year.

Larger schools, but not necessarily gigantic ones, offered several educational advantages; better trained teachers, fuller range of courses, better facilities. The arguments by Conant and other proponents of larger schools were essentially threshold arguments. Schools must attain a certain size in order to be effective, efficient or equitable - or possibly all three. Not everyone bought this argument, however. The nation did not continue consolidating until Conant's threshold was met.

Opposition Argument

People have fought school district consolidation on several grounds. Rural residents wonder if a school in town will really serve the needs of the farm children.   Financial considerations have also had many to oppose school district consolidation. Iowa farmers, for example, did not wish to pay to educate town children when ad valorem taxes totally financed education in that state (Lancelot 1947: 66).   Even proponents of consolidation often admit that taxes seldom go down and often go up after consolidation, though they often argue that the product is sufficiently superior to justify any marginal added expense.   Just as with Welsh local government, scale efficiencies tend to be lost to increasing the level of services - or more generous compensation to staff (Sher and Tompkins 1977: 46).

Transportation costs are another focus for opposition. Nowadays these costs are largely in terms of money and time. If distances are significant some children will pay a heavy price in time consumed going to and from school.   Honey and Kohler (1977) have shown that even a modest threshold of 400 pupils for an entire school district would force an additional 20-mile trip for some children in Iowa.   Opposition because of distance has been found in an analysis of referenda on school consolidation. The farther people are from school in a proposed consolidation, the less likely they are to support the measure (Shelley 1977: 11).

Loss of a school, especially a high school, may present a real blow to a town.   First, the school may be an important element in the town's economic base.   Possibly more important would be the blow to the town's pride. Schools, again particularly high schools, tend to be the focus of great attention in the U.S.A., especially in small towns. Many residents would doubtlessly fight such a threat to local identity.

Finally, school district consolidation is opposed on the very grounds of its major support - quality.   Sher (1977) argues strongly that the evidence for quality of modestly sized schools is as great as that for large schools, so consolidation should be stopped before accomplishments become catastrophes.

Reorganisation Process

Whatever the arguments for school district reorganisation, the fact is that districts have consolidated.   The way they have done so is important because it has generated the products now in existence. In fact, the reduction of school districts has taken place under 50 sets of rules - one per state - and the rules themselves have been changed many times. Despite the rules and any other differences, however, extensive reorganisation has taken place. Between 1930 and 1961 every state eliminated at least 60 per cent of its districts and 80 per cent of the states eliminated more than 90 per cent of their districts (American Association of School Administrators 1962: 10).

Some states, particularly in the South, have adopted county school systems. Most, however, have left the change up to local decision-makers, especially the voters. The states have tended to use a carrot and stick approach, offering incentives to reach certain thresholds. They have also acted as brokers, matching likely candidates for mergers.

Mergers are exactly what most reorganisations have been. Two or more districts simply become one. Typically the new district would have only a single high school but usually more than one elementary school, at least at first. Given the ad hoc nature of the mergers - particularly the very real chance that the voters in one or more candidates for merger will reject the issue - the products often have highly irregular boundaries. A small district may be left with no potential partners except a couple of very large neighbours. The distribution will be the product of many local decisions and generally bears little relationship to what would have occurred had a system of districts been planned.

Consequences of Change

Unquestionably one major change has been a significant centralisation of authority. Where small rural areas used to act as a unit, much more sizable units are now operating. As the area of daily contact has grown with contemporary transportation, so has the school district grown (Shupe 1947: 42). The changing scale of authority sometimes produces unforeseen conflict. Recently one Iowa school district has been torn over a school closing controversy. Several districts merged to form the present one years ago. The largest town has the district high school but the smaller towns retained their elementary schools. Now, with enrolment declining and the district facing a financial crisis, the district officers have decided to close one of the small town schools. The local residents stated a preference towards leaving that particular district and joining another that would allow them retain their school. Local control has indeed been lost.

Another shift in power has been from rural to urban areas. The towns, with their greater voting strength, can usually win a political fight. Some argue that this is simply the culmination of more than 100 years of urbanising rural schools (Rosenfeld and Sher 1977: 11-42).

Power has also shifted from the population as a whole to professionals (Great Plains Project 1968: 1-12; Rosenfeld and Sher 1977: 41; Association American School Administrators 1958: 59-74). Control over schools has become less a question of what the people in a locality want and more a question of what various education experts think the people should have or what the experts think need to be provided so that the people can have the outcome they seek.

A number of geographical consequences can be seen from a century of school reorganisation. As the scale of social and economic life has changed schools have had to change, both in content and in spatial organisation. The change has not been uniform, however. It has depended on state laws as well as local conditions. Varying densities, for example, have permitted some areas to achieve Conant's threshold while for other areas the costs in accessibility would have been too great.

Aside from the athletic exploits of a school's team, emotional attachments to the past do not appear to be very significant. Continuity is seen more through the additive procedure for change than in retained identities. School jurisdictions have proven to be flexible, giving the public education system the versatility it has needed to survive the very great changes in its situation.

CONCLUSION

The jurisdictional organisation of space is indeed a major dimension of societal territoriality. Many people gain a sense of identity; all gain public services - and a tax burden. As society changes jurisdictions tend to follow but usually with a significant lag. Pressures from flexibility, versatility fly against pressures for continuity, history. The status quo vests power in specific ways and those holding this power fight to keep it. Only in a society fixed in all relationships would one expect a constant order of jurisdictional space, however. Such change is inevitable - and the geographical distribution of power will change along with it.

REFERENCES

AMERICAN ASSOCIATION OF SCHOOL ADMINISTRATORS [1958]
  School District Organization, (AASA, Washington, D.C.).
AMERICAN ASSOCIATION OF SCHOOL ADMINISTRATORS [1962]
  School District Organization: Journey That Must Not End,
  (AASA, Washington, D.C.)

ARCHER, J.C. [1981]
   'Public choice paradigms in political geography', in A.D.
   Burnett and P.J. Taylor (eds.), Political Studies From
   Spatial Perspectives, (Wiley, Chichester), 73-90.
CLARK, G.L. [1981]
   'Democracy and the capitalist State: towards a critique
   of the Tiebout hypothesis', in A.D. Burnett and P.J. Taylor
   (eds.), Political Studies, 111-130.
CONANT, J.B. [1959]
   The American High School Today: A First Report to Interested
   Citizens, (McGraw-Hill, New York).
CUSHMAN, M.L. (ed.) [1947]
   Proceedings of the Third Annual Iowa State College Conference
   of Problems and Procedures of School District Reorganization
   in Iowa, (Iowa State College, Ames).
DEAR, M. [1981]
   'A theory of the local state', in A.D. Burnett and P.J.
   Taylor (eds.), Political Studies, 183-200.
FESLER, J.F. [1949]
   Area and Administration (Alabama University Press,
   Tuscaloosa).
FREEMAN, O. [1968]
   'Rural despair and a rural renaissance', in Planning for
   School District Organization, (The Great Plains School
   District Organization Project, Lincoln).
FREEMAN, T.W. [1968]
   Geography and Regional Administration, (Hutchinson University
   Library, London).
GREAT PLAINS SCHOOL DISTRICT ORGANIZATION PROJECT [1968]
   Guidelines for School District Organization, (Lincoln,
   Nebraska).
HONEY, R.D. [1981]
   'Alternative approaches to local government change', in
   A.D. Burnett and P.J. Taylor (eds.), Political Studies,
   245-74.
HONEY, R.D. and J. KOHLER [1977]
   Distance Effect of School District Reorganization, (State
   of Iowa Department of Public Instruction, Des Moines).
LANCELOT, W.H. [1947]
   'Possible influences of Iowa's State Aid Program', in W.L.
   Cushman (ed.), Proceedings, 66-70.
LEA, A.C. [1979]
   'Welfare theory, public goods and public facility location',
   Geographical Analysis, 11, 217-39.
LOCAL GOVERNMENT COMMISSION FOR WALES [1962]
   Report and Proposals for Wales, (HMSO, London).
LOCAL GOVERNMENT BILL [1972]
   (HMSO, London).

LUKERMAN, F. [1961]

'The concept of location in classical geography', Annals of the Association of American Geographers, 51, 194-210.

MASSAM, B. [1975]

Location and Space in Social Administration, (Edward Arnold, London).

NORTON, A. [1979]

'Britain: England and Wales', in D.C. Rowat (ed.), International Handbook on Local Government Reorganization, (Greenwood Press, Westport), 261-74.

REFORM OF LOCAL GOVERNMENT IN SCOTLAND [1971]

(HMSO, London).

REFORM OF LOCAL GOVERNMENT IN WALES [1971]

(HMSO, London).

RHODES, R.A.W. [1979]

'Developed countries', in D.C. Rowat (ed.), International Handbook, 563-81.

ROSENFELD, S.A. and J.P. SHER [1977]

'The urbanization of rural schools, 1840-1970', in J.P. Sher (ed.), Education in Rural America: A Reassessment of Conventional Wisdom, (Westview Press, Boulder), 11-42.

ROWAT, D.C. (ed.) [1980]

International Handbook on Local Government Reorganization: Contemporary Developments, (Greenwood Press, Westport).

ROYAL COMMISSION ON LOCAL GOVERNMENT IN GREATER LONDON, 1957-1960 [1960]

Report, (HMSO, London).

SCHMANDT, H.J. et al. [1961]

Metropolitan Reform in St. Louis: A Case Study, (Holt, Rinehart and Winston, New York).

SHARPE, L.J. [1979]

Decentralist Trends in Western Democracies, (Sage, London).

SHELLEY, F.M. [1977]

The Effects of Distance on Voter Support for School Decentralization in Rural Iowa, (Unpublished paper, Department of Geography, University of Iowa).

SHER, J.P. [1979]

Education in Rural America: A Reassessment of Conventional Wisdom, (Westview Press, Boulder).

SHER, J.P. and R. TOMPKINS [1979]

'Economy, efficiency and equality: the myth of rural school and district consolidation', in J.P. Sher (ed.), Education in Rural America, 43-77.

SHUPE, L.E. [1947]

'Lesson from the history of school transportation' in M.L. Cushman (ed.), Proceedings, 42-46.

SMALLWOOD, F. [1965]
  The Politics of Metropolitan Reform, (Bobbs-Merrill, Indianapolis).
SOJA, E.W. [1971]
  The Political Organization of Space, Resource Paper #8, (Association of American Geographers, Washington, D.C.).
THOMAS, P. [1975]
  Personal interview with former Secretary of State for Wales, 3 July 1975.

Chapter 19
INTERGOVERNMENTAL RELATIONSHIPS AND THE TERRITORIAL STRUCTURE OF FEDERAL STATES

Ronan Paddison

INTRODUCTION

Reorganising the territorial networks of governments into which the state is divided is invariably a sensitive issue, what Wood (1974) called a political 'hot potato'. Usually encountering the opposition of the localities themselves, political influences at the centre may also militate against reorganisation. National politicians with a local constituency vote to cultivate will be mindful of local interests. This chapter examines the problems of implementing reorganisation in Australia as they have arisen within the federation and, more particularly, as they arose following the attempts at recasting the map by the short- lived Labor Government of 1972-5. Two related propositions are examined. First, that attempts to alter the structure <u>nationally</u> within a federation are problematic precisely because of the federal make-up of the state and, second, that any such attempts will need to find some compromise between the different levels of government party to the change. Failing such compromise, any new structure will probably be short-lived.

In contrast to the unitary state characterised by a politically powerful centre and in which subnational governments are constitutionally subordinate to the centre, federal polities comprise a varying number of sovereign component states or provinces, in theory coequal to the federal or national government. Below the states, and often constitutionally subordinate to them, is the network of local governments. It is because of the existence of the component states that any attempt at radical jurisdictional change nationally within a federal state is the more problematic.

In the classic 'federations' local government is a state or provincial concern. The national government is thus in a relatively weak position to impose, or even encourage, territorial reorganisation over the country as a whole, as has been proposed and implemented within several states in the developed world (Rowat 1980).

This is true, in particular, of Australia, where policies of the Labor Government were directed at the restructuring of local government. In the constitutional division of powers, this was a field within the sphere of competence of the state, and not the federal, government. During the Labor administration, the federal government's policies were directed at local government reorganisation through the formation of new regional bodies. Ideally, it was hoped that the definition and functioning of these bodies would be supported by the six component states.

However, national policies inspired by the federal government, will meet the opposition of the subnational bodies, to a greater or lesser degree, because of their possible centralising effect. In the federal state, decentralisation (or noncentralisation) is an important element of the national political culture.

In the recent Australian experience, reorganisation is a misnomer. Reorganisation is normally defined as a process involving both the death and birth of new territorial institutions. However, the policies of the Labor Party sought a reallocation of functions between the three existing levels of government, federal, state and local, adding the new regional bodies as a fourth tier between the state and local governments. At the local level, these were not actually replacing any existing authorities in the short term. In the longer term, the federal government's hopes were that regionalisation would encourage the voluntary amalgamation of local governments. This was to be expected after they had seen the benefits of planning and managing the delivery of services over a wider area.

To the localities, as well as the states, regionalisation was seen as a potential threat to their own power base, so that intergovernmental conflict was perhaps inevitable.

IDEOLOGY AND THE FEDERAL STRUCTURE

Federalism, as the constitutional lawyer A.V. Dicey argued, 'rests on the psychology of the peoples of the political units desiring union without unity' (Dicey 1885; 602). It is an expressly territorial mode of government embracing demands for national union and the maintenance of a political commitment to the constituent units of the federation, the states or provinces. Among those federations which have been derived through the voluntary amalgamation of the component states the commitment to both levels is enshrined in the constitution. Both the federal and state governments are considered sovereign, co-equal partners. Although in practice the national government may become 'first among equals' the continuing relationships between the two levels of government must be such as to respect the federal bargain. Thus, in the constitution, the rights and responsibilities of the different levels of government are spelt out. Infringement by one within the sphere of competence of the other will be an obvious source of conflict initiating accusations that the spirit of the federation has been broken.

Australia is a federation which was derived 'from below'. Established in 1901 through the voluntary union of the six states, the constitution provided a formal division of powers between the Commonwealth (federal) and state governments. It stipulated in detail those powers given to the Commonwealth government leaving the 'residuary powers' to the states. Significantly, for the recent Labor initiatives, local government was not a signatory to the constitutional compact and was left as a purely state concern.

In Australia, both before and after federation, federalism has been supported and opposed by various groups. From its inception the advantages of federalism have been contested. In part the argument centred around the gains and losses of the interests of the separate colonies by federalising (Dikshit 1975). Arguably, a more enduring conflict over a federal or more centralised form of government has arisen between the two major national political parties, Liberal and Labor. In this sense the advantages and disadvantages of federalism have become polarised around partisan and class interests.

To the Liberal Party (as also to its coalition partner, the Country Party) the virtues of federalism are those which are commonly considered as the advantages for this type of government (Wheare 1963). Fundamentally, federalism is to be justified on philosophical grounds; as de Tocqueville might have said:

> ' by preventing the concentration of power
> in a few hands. . . it provides a guarantee
> of political and individual freedom' (Liberal
> and National Country Parties Federal Policy
> Statement 1975).

Traditionally the Labor Party has been antipathetic to a federated Australia. In the eyes of Labor, federalism was seen to favour business commercial, regional or sectional interests more than wider, national interests or those of the urban proletariat. In the early twentieth century the Australian Labor Party (ALP) had already adopted unification as official policy essentially because this would provide the means by which its socialisation policies could be implemented nationally. Besides meeting the opposition of the states generally, it was a policy which encountered resistance from within the party's own ranks, for given the decentralised organisation of the party it ran against the State Labor parties' interests (Emy 1978).

By the time the ALP had launched its regionalist policies the party had formally abandoned unification. Nevertheless, it is within this tradition of anti-federalism that the broad objectives of Labor's regionalism programme should be seen. They were seen as such by their opponents. Gough Whitlam, the ALP leader, had shown a continuing interest in the question of appropriate governmental structures. His antipathy to federalism was expressed on several occasions.

While unification had formally been dropped by the party – it was both a vote-loser and unrealistic in that the states had become considerably more entrenched since the policy was first adopted – Whitlam's thoughts on the 'ideal map', little different from the proposals discussed earlier in the century, were to provide his opposition with ammunition. Talk of decentralisation, it was argued, was rhetoric and Labor's real intentions were the opposite.

One of the reasons which help to explain the anti- federalist strain within the ALP was simply that a more centralised state would help ensure that the policies of the centre would prevail. Put differently, in the federated state it is not inconceivable that the resistance of one or more regional (provincial) governments could impede the implementation of federal policies. Inevitably the implementation of national policies would rely upon the cooperation of the states (and also, often, of local government) and although since federation, in common with the pattern in other federal countries (Sawer 1969), the trend has been towards increasing cooperation between federal and subnational governments. There has also been an increasing dominance of the national government, especially in fiscal matters and the states could (and sometimes did) act as a brake on policies the implementation of which was to be national.

As the Labor government was to discover, intergovernmental cooperation in practice can actually mean conflict based on the central values on which the federal bargain is struck, the simultaneous desire for national union and regional diversity. Federal-state conflict often arises because of the different emphasis placed on these values by the two levels of government. Equally, federal and state politicians need to cultivate their own electorate and even where the same party holds office at both levels, electoral concerns can bring about a conflict of interests. Thus in South Australia the state Labor Party had to formally dissociate from the regionalist policies of the ALP in order to retain office in the 1975 elections (Figure 19.1).

TERRITORIAL OSSIFICATION

Even though the Labor government's espousal of substate regionalism was to be a more diluted form of territorial change than Labor politicians might have subscribed to in an ideal world, the 'vested interests' of which Whitlam spoke were still to be an obstacle to change. (The policies did not involve, as pointed out earlier, any territorial reorganisation in the usual sense.) Territorially these vested interests had become a well-established tradition within Australia; indeed, given the small number of boundary changes that have occurred within the last 100 years, most of which were limited to local government, it would not be unfair to describe the structure as having ossified (Power and Wettenhall 1976). This hardening of the

Figure 19.1:   Alternative Territorial Systems - Australia

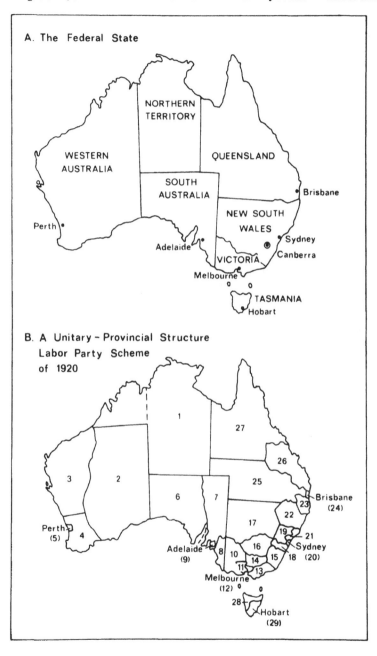

A. The Federal State

NORTHERN TERRITORY

WESTERN AUSTRALIA

QUEENSLAND

SOUTH AUSTRALIA

Brisbane

NEW SOUTH WALES

Perth

Adelaide

Sydney

Canberra

VICTORIA

Melbourne

TASMANIA

Hobart

B. A Unitary - Provincial Structure
Labor Party Scheme
of 1920

1

27

26

3

2

25

6

7

Brisbane
(24)

23

22

Perth
(5)

4

17

19

21

Adelaide
(9)

8

10

16

15

Sydney
18   (20)

14

11

13

Melbourne
(12)

28

Hobart
(29)

boundaries had become prominent at both state and local levels more so in the former than the latter. Its implications were to be important for the federal Labor Government's attempts to encourage an extra, intermediate territorial body.

The Labor Government's recognition of the states, explicit in the formal abandonment of unification, represented the acceptance of the politically inevitable; it meant also that the boundaries of the states would have to be respected, regardless of any anachronisms they might pose. One of the defining characteristics of federalism is that the territorial limits of the component states should be inviolable except for the acquiescence of the states themselves (Duchacek 1970). Save for possible minor boundary changes it is unlikely that any such body will vote for its own dismemberment. The rightness of this stability can also be supported. Were such changes possible unilaterally, presumably by legislative sanction of the central authority rather than by local secession the federal character of the state would be undermined. While the inviolability of the state boundaries is common to the federal countries, it is obviously more pronounced among the federations built 'from below'. Both cases in which state boundaries have been altered comparatively easily lack this organic evolution. In West Germany federalism was established within a somewhat unsatisfactory territorital framework, the lander to a large extent following the post-1945 Allied Occupation zones; in India the federal government can initiate the process of territorial reform. This, together with other facets of the constitution, raise doubts as to whether India can be categorised as truly federal.

Given the origins of the Australian federation it is not surprising to learn that the existing states have successfully resisted attempts for their subdivision since 1901. These attempts became most closely associated with the 'New State' movements, which were most active in New South Wales and Queensland. New staters argued that the existing states were too large and were politically dominated by geographical/sectional interests. Often located on the margins of the existing state the New State movements represented centre-periphery conflict. It was in the resistance to the formation of New States that the constitutional strength of the states became apparent (Whebell 1973). Some of the movements commanded sufficient local legitimacy to 'force' the holding of a referendum - notably New England in northern New South Wales (Woolmington 1966), though by gerrymandering the area in order to include populations that are known to be unsympathetic to secession, the position of the state was thus reasonably assured.

One of the features highlighted by the New State movements and of relevance to recent Labor attempts to reorganise the territorial map was the relative weakness of the federal government. The subdivision of the states into new governmental units - in this case new states - was effectively left to the states themselves. Indeed, an attempt in the 1920s by the New England movement to draw the federal government into the problem resulted in a State's Right response from New South Wales.

If the states have been successful in maintaining the territorial status quo, so too has local government, and given its weaker constitutional position, the ability of local government to resist change is all the more noteworthy. This is not to say that territorial change has not been discussed. Each of the states, with the exception of Queensland, has commissioned at least one inquiry to examine the question of reorganisation, the tenor of which can be gauged from the following quote in the New South Wales report

> ' . . . Local government suffers from the existence of too many small uneconomic areas, resulting in the fragmentation of authority, unnecessary duplication of assets, the underutilisation of plant, equipment and human resources and an inability to provide the varied kinds of expertise required by local councils in the modern world' (Barnett Report 1974: 52).

To the states, then, the problem was essentially one of size and of the achievement of rationalist objectives, notably greater efficiency and economy. To the local councils, on the other hand, the problem was seen as one of providing more financial aid so as to attain a better level of service.

The reasons for the failure of local government are important to our argument. The prime reason is their success in mobilising support for the status quo by legally challenging the inquiries, notably in Victoria and Tasmania (Rawlinson 1975). Reform has also lacked powerful lobby support in the state legislatures. Finally, the states have been equivocal over the desirability of reform, for where local government, as presently structured, is relatively the weak partner, a smaller number of larger jurisdictions might serve to alter the balance of power.

LABOR'S REGIONAL POLICIES

The return of the Labor Party to power in 1972, ending 23 years of Liberal-Country Party rule, promised to mark a turning point in Australian politics. In his major policy speech of December 1972 Whitlam outlined the three main aims of ALP policies: 'to promote equality, to involve people in the decision- making processes and to liberate and uplift the horizons of the Australian people'. Behind the rhetoric there are two facets underlying these aims which were to be important from a structural viewpoint.

Firstly, the equality had been redefined by Whitlam as one of opportunity and not of income; thus inequalities were locational and closely allied to how real income was distributed within the city. These locational inequalities, associated particularly with the distribution of public services were most severe in the outer suburbs of the capital cities and it was the promise of improvements in these areas that had contributed to a change in voter allegiances and the return of the Labor Party to power. Federal aid was to be directed particularly to the metropolitan areas so as to redress the 'private affluence and public squalour' from which the cities suffered.

Secondly, as these inequalities were more marked between 'regions' than they were between states, remedial action would be more appropriately channelled through regionally-defined bodies than through the states. Put another way, if the federal government was to provide financial assistance to reduce these inequalities, then it appeared that the most effective implementation of the pogrammes by a new level of territorial authority, the substate region, embracing a collection of local governments. The states were too few and internally heterogeneous and the individual local governments too numerous to be effective.

The preference for regional bodies was to serve two other objectives of ALP policy. They would provide the means by which greater participation in decision-making could be incorporated. However, with the partial exception of the Australian Assistance Plan, a social welfare programme, this objective was poorly articulated and was one of the least successful of the regionalist objectives.

Thirdly, local government was to feature prominently within regionalisation. The regions represented one means by which the status of the 'third tier' of government was to be upgraded. As Whitlam strove to impress, so many of the services affecting life-chances - health, housing, transport, child care and so forth - depended on having strong local government; but the states had been 'indifferent' to the problems of the local councils and he argued that the local councils should have 'proper access to the nation's financial resources' (Whitlam 1974: 161). This was to be achieved by extending the Grants Commission scheme by which the poorer states had enjoyed 'topping up' payments so that public service provision approached the standard of that provided in the more affluent states, to local government, and by which federal grants would be provided to 'approved regional organisations'.

Two federal departments dominated in regionalising schemes: the newly formed Department of Urban and Regional Development (DURD) and the Social Welfare Commission (SWC). By the end of 1973 DURD had divided the country into 68 regions in each of which a Regional Organisation of Councils (ROC) was to be established through which the Grants Commission equalisation payments would be made. Subsequently it was through the ROCs that Area Improvement Programme payments would be made. These payments were channelled mainly towards the metropolitan areas and particularly to the Western regions of Melbourne and Sydney

where infrastructural and community facilities were most lacking.
The SWC, through its innovatory social welfare programme, the
Australian Assistance Plan, encouraged the formation of regional
councils through which federal aid, promoting social welfare was
directed.

DEFINING THE REGIONS

While within Canberra there was much talk of regionalisation
and its advantages, much groundwork needed to be covered before
a national network of regional bodies could be implemented.
Obviously an initial task was delimiting the regions on the
ground, a task which DURD contracted out to a group of Melbourne
geographers (Logan, Maher, McKay and Humphreys 1975). Their
'solution' was based on the definition of a set of nodal regions
in the non-metropolitan areas, identified on the basis of a
cluster analysis of telephone traffic between 324 exchange areas.
While preferring a nodal solution within the non-metropolitan
areas because of the alleged community of interest of town and
hinterland and the provision of a rationale on which spatial
investment policies could be developed, in the capital cities
the regions were defined more on the basis of their socio-
economic homogeneity. This would help define areas within the
metropolitan complexes which were relatively disadvantaged in
terms of their social and infrastructural equipment, and defined,
in other words, likely problem areas.

Having developed its own option, it was then necessary for
the federal government to seek consultation with the states as
to the issue of regional definition. This process was complicated
by the fact that most of the states had already embarked on the
regionalising of public service activities. In Queensland and
New South Wales, particularly, regionalisation of state
administration was well developed. Each had regional advisory
bodies coordinating physical planning and other services. In
New South Wales this had developed to the point that a set of
'Standard Regions' had been defined to which state department
field services were meant to (and mostly did) subscribe. In
contrast, Western and South Australia had only begun to think
in regional terms.

Regardless of how far state regional thinking had progressed
the federal DURD regions were tailored more often than not to
state preferences. The federal department insisted that no
individual local governments be split between two regions, for
this would render impossible the task of the Grants Commission
programme.

Ideally DURD sought a national set of regions defined on
the basis of a common set of criteria, but preexisting state
regionalisation schemes and the antipathy of the states to
'federal centralism' militated against the likelihood of such
an outcome. Nowhere was this problem more apparent than in

Queensland and particularly within the metropolitan area centred in Brisbane. While DURD would have preferred to subdivide the Moreton Bay region - even though Brisbane, unlike the other state capitals, was under unified local government control - the state Coordinator-General's definition treated the area as a single unit. The antipathy of the state governments to the ALP's centralism was strongest in Queensland as the state was controlled by a right-wing party. Federal-state conflict during the time of the Whitlam government was by no means restricted to the regionalist policies (other federal 'incursions' included attempts to control mineral extraction and the treatment of Aborigines), but the state refused to negotiate over issues like boundaries which were considered purely a state affair.

It is not unlikely that DURD officials 'soft pedalled' the boundary issue because state and local cooperation was considered important for the policies to be pursued through regionalism and were not worth endangering for the sake of structural preferences. But it is evident that the implementation of a national system of regions needed to compromise with state and local priorities. Another problem was added by the conflict between the federal regionalising departments over the issue of suitable regional boundaries. The SWC, through the Australian Assistance Plan, sought regions smaller than DURD in the hope that this would encourage participation which constituted a more important element within their programme. To DURD the regions defined by SWC were to constitute a further problem in establishing a common set of national regions through which federal policies could be implemented and upon which a form of 'regional consciousness' could be built.

REGIONAL GOVERNMENT OR REGIONAL ADMINISTRATION?

More contentious and more problematic than the issue of boundary definition was the question of the status of the regional bodies. In particular, were they an attempt by the Commonwealth Government to implant a fourth tier of government or were the bodies merely to have an administrative role unlikely to threaten the power bases of the existing state or local bodies? While the former would be unpalatable to both the states and local government the latter would undermine the viability of regionalism as a political and enduring force. Lacking local legitimacy as administrative bodies such as that gained through direct elections the criticism was that they were merely vehicles through which national government would ensure the implementation of its own policies in the localities. In other words, regionalism was to be seen as a centralising strategy, rather than, as Whitlam had originallly suggested, a real attempt at decentralisation.

If Whitlam had started with the intention of regional government this was abandoned once in office (Power and Wettenhall 1976). The position was spelled out by DURD, the arch- regionalising federal department, which pointed out that regionalism did not equal rational government.

Apart from being captive within an ossified structure, in which federally inspired regionalism was constrained by the territorial realities, another reason explaining the retreat from regional government was the difficulty in gaining consensus within the federal government as to the objectives of regionalism. The conflict between the two major regionalising departments in terms of the boundary definition issue has been noted. More significantly, the two departments had differing ideas as to who should be represented on the regional bodies; while through the ROCs, the DURD emphasised representation by local government, the SWC played down the role of formal bodies (Paddison 1977). Consensus being difficult, it proved impossible to coordinate federal regionalist thrust within the regions. Function dominated over area as the organising principle so that regionalism became a package of loosely coordinated policies rather than an integrated set of programmes giving clear expression to the regional dimension.

While the problems of coordination between the regionalising departments in the federal government blunted the regional thrust, and while the advocacy of regional government had been lost, DURD had its own aspirations as to the direction in which regionalism should lead. Careful not to mention local government, DURD argued that in the 'very long term' the ROCs could become the 'regional arm' of elected local government, responsible for regional planning and coordination. Given the initial encouragement to form themselves into regional groupings, as the Grants Commission scheme had done, the local councils would come to see increasing value in inter-area cooperation to the point that they would be willing to delegate planning powers. Increasingly, regionalism was seen in administrative terms, as a means by which to overcome the problems of jurisdictional fragmentation and as the vehicles through which federal, state and local governments could cordinate their activities. If the longer term aims were not to be realised because of the premature demise of the Labor Government, the beginning of such coordination was apparent in some areas, notably the Area Improvement Programmes in the metropolitan centres. It was in such areas that the federal grants were to be of considerable significance in making inroads to the improvement of infrastructural services (Figure 19.2). In those regions covering the relatively deprived western suburbs of Sydney and Melbourne the benefits of regionalisation had begun to be evident and with sufficient time it would not have been unlikely that the regional bodies would have come to enjoy delegated powers along the lines suggested by DURD.

The basic problem confronted by the short-lived Labor regionalist policies was the antipathy of the states to what they considered to be 'creeping centralism'. Had they been brought more directly into Labor's regionalism from the outset, the initiatives might have fallen on more fertile ground.

However, given the almost 'natural' antipathy between the federal government and some of the states this would have been difficult. Such an approach would at least give recognition to the territorial realities of Australian government.

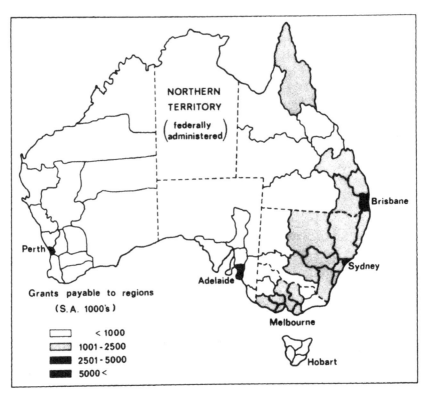

Figure 19:2 Australian Government Regions and Commonwealth Grants Commission's Grants, 1975.

CONCLUSIONS

This work has outlined some of the problems which have been encountered in successive attempts to modernise the jurisdictional structure of the state. In Australia territorial change has been limited to piecemeal alteration of the local government map.

In comparison with the unitary state the distinctive feature of the federation is the interposition of the component states between national and local governments. These are intermediate-level governments which both constitutionally and politically are powerful and indissoluble elements of the institutional makeup. Altering the boundaries of these intermediate- level governments in the federal state is unlikely and in this respect Australia shares the experience of the other federations. But the Australian experience also highlights one other way in which territorial change is problematic in the federal state, where any attempt by the national government to establish territorial innovation will of necessity need the cooperation of the states. Attempting to 'bypass the states', a commonly held interpretation of the regionalist policies of the Whitlam government, would only exacerbate federal-state conflict and raise the spectre of state's rights.

The very processes by which the federal state in its 'classical' form, in contrast to its unitary counterpart, was established also impinge on the question examined here. In the federal state the process of formation, the voluntary amalgamation of previously separate units, continues to have a political significance founded upon the constitutional recognition given to the states as well as to the federal government. Typically, in the federal polity there will be a strong anti-centralist tradition so that attempts to implement national reforms by a federal government will be criticised for their centralising effects. In Australia this tradition is, if anything, sharper where the ALP is concerned because of its popular perception as the 'centralising' party.

REFERENCES

BARNETT REPORT ON LOCAL GOVERNMENT [1974]
     (Government Printer, Sydney).
DICEY, A.V. [1885]
     Introduction to the Study of the Law of the Constitution,
     (London).
DIKSHIT, R. [1975]
     The Political Geography of Federalism, (Macmillan, New
     Delhi).
DUCHACEK, I.D. [1970]
     Comparative Federalism, (Holt, Rinehart and Winston, New
     York).
DURD [1973]
     Regions, (Parliamentary Paper, No.246, (AGPS,
     Canberra).
DURD [1975]
     Australian Government Regional Boundaries, (AGPS,
     Canberra).
ELLIS, U.R. [1933]
     New Australian States, (The Endeavour Press, Sydney).
EMY, H.V. [1978]
     The Politics of Australian Democracy, (Macmillan,
     Melbourne).

LIBERAL AND NATIONAL COUNTRY PARTIES [1975]
Federal Policy Statement

LOGAN, M.I., C.A. MAHER, J. McKAY and J.S. HUMPHREYS [1975]
Urban and Regional Australia; Analysis and Policy Issues, (Sorrett, Melbourne).

PADDISON, R. [1977]
The Political Geography of Regionalism, (University of New England, Armidale).

POWER, J.M. and R.L. WETTENHALL [1976]
Regional Government versus Regional Programs, (Mimeo Paper, Canberra).

RAWLINSON, M. [1975]
'Administering Local Government reform in Australia: the State Experience', in R. Wettenhall and M. Painter (eds.), The First Thousand Days of Labor, (CAE, Canberra), Volume 2, 67-82.

ROWAT, D.C. [1980]
International Handbook on Local Government Reorganization, (Aldwych Press, London).

SAWER, G. [1969]
Modern Federalism, (Pitman, Carlton).

WHEARE, K.C. [1963]
Federal Government, Fourth edition, (Oxford University Press, London).

WHEBELL, C.F.J. [1973]
'A model of territorial separatism', Proceedings of the Association of American Geographers, 5, 295-98.

WHITLAM, E.G. [1974]
'New directions in local government', Australian Government Digest, 2(1), 161-66.

WOOD, B. [1974]
The Process of Local Government Reform 1966-74, (George Allen and Unwin, London).

WOOLMINGTON, F.A. [1966]
A Spatial Approach to the Measurement of Support for the Separatist Movement in Northern New South Wales, (University of New England, Armidale).

258

Chapter 20
BOUNDARIES AS BARRIERS — THE PREDICAMENT OF LAND-LOCKED COUNTRIES

Hendrik-Jan A. Reitsma

'We cannot obliterate the harsh facts of
history and (political) geography . . . for
obvious reasons, Botswana must maintain
diplomatic contacts with South Africa.'
(Prime Minister Sir Seretse Khama of Botswana,
1970)

INTRODUCTION

Political boundaries, and especially international boundaries,
are known to act as barriers. Because they tend to interfere
with 'free' spatial interaction in a variety of ways, it is not
uncommon that things on one side of them differ markedly from
those on the other side. Such cross-boundary contrasts might be
cultural in nature or political, economic or social. They may
be quite insignificant, such as the differences in postage stamps
or the brands of cigarettes, or they might be more important,
such as differences in nationalistic feelings, political ideology,
degree of industrialisation, unemployment rates, ethno- demographic
composition, or the quality of public services. Even the landscape
can be surprisingly different on opposite sides of a boundary;
densely populated and urbanised versus sparsely settled and
rural, as an example. Elsewhere, small private farms with mixed
farming may constrast with large grain-producing state farms.
Even along the border between Canada and the U.S.A., two countries
which are so alike in so many ways, crop and livestock combinations
north of the border differ from those to the south of it (Reitsma
1969; 1971; 1972).

Many more cross-boundary contrasts can be listed but that
is not the purpose here. It should be stressed, however, that
all these differences exist because of the barrier effects that
international boundaries normally have.

Barrier effects are particularly great when people purposely
use boundaries to stop all movement of goods and humans between
contiguous political entities. Examples of such closed borders
are the Berlin Wall and the so-called Iron and Bamboo Curtains.

259

Whereas Western Europe has only one closed border - between Spain and Gibraltar (re-opened in December 1982) - the African continent has many boundaries that have been closed for shorter or longer periods of time since the early 1960s.

Because movement (e.g. spatial diffusion) is a prerequisite for change, growth and progress, interference with movement constitutes a hindrance to development. This pertains not only to the movement of goods, or commodity trade, but also to that of persons, ideas, knowledge, skills, technology and capital. Lack of exchange, like isolation and inaccessibility, will sooner or later result in stagnation and decline, in cultural and economic backwaters. No developed country or thriving civilisation is an island unto itself.

UNDERDEVELOPMENT AND GEOGRAPHY

In recent years, much has been written about the problem of underdevelopment, regardless of how the term is defined. Especially (political) economists have given it a great deal of attention. But also many (economic) geographers have made contributions to the study of the development problematique. As yet, however, much of their (theoretical) work is not very geographical in nature, probably because most of them did not climb on to the development bandwagon until relatively late. Despite their having adopted such concepts as 'core', 'periphery' and 'growth pole', not a few of their empirical studies and theoretical discussions are essentially economic in nature, with more attention paid to unequal exchange, foreign investment, surplus extraction and integration into the (world) capitalist system than to such factors as relative location, geopolitical fragmentation and distance. And like the political economists before them, they are often more concerned with (dependency) relations between Third World countries on the one hand and developed, capitalist states on the other, than with (dependency) relations among Third World countries, such as those between Nepal and India, Bolivia and Brazil, or Uganda and Kenya.

THE LEAST DEVELOPED COUNTRIES AND DEPENDENCE THEORY

Maps of the world's least developed (or poorest) countries show patterns that are most interesting when considered from a purely geographical point of view. No matter how 'least developed' or 'poorest' is defined and regardless of whether the maps were made ten years ago or more recently, they invariably reveal that many of the least developed states happen to be land-locked countries (LLCs). Although LLCs make up no more than 20 per cent of all states, they account for 50 per cent of the poorest countries.[2] This over-representation of LLCs is all the more remarkable when it is realised that one out of every four LLCs

260

is located in Europe, none of which is generally considered to be underdeveloped. In other words, the overwhelming majority of extra-European LLCs belong to the group of poorest countries. The only clear exceptions are Zambia and Zimbabwe in Africa[3], Mongolia in Asia, and Bolivia and Paraguay in South America, even though the latter two are the least developed countries on the South American continent.

This high degree of correlation has significance, so the question we are faced with is: How do we explain that nearly every non-European LLC is plagued with such an extremely low standard of living?

According to Dependencia theory, LLCs can be expected to be less underdeveloped than coastal states (CSs). Because of their relative inaccessibility, they should be, and in fact are, comparatively unappealing to profit-seeking enterprises. Consequently, they have not attracted much investment, (except for countries such as copper-rich Zambia), know little capitalist penetration and are only marginally integrated into the capitalist world-economy. According to the Dependencia thesis, LLCs should be less dependent than CSs on the developed countries of the West and should also be less exploited. To use an expression of Frank (1978), the 'dependent development of underdevelopment' cannot have progressed as far in the Third World's LLCs as it has in the Third World's CSs. The dependency view, then, would have us believe that the LLCs have less poverty and less socio-economic and spatial inequality (or polarisation) than the more accessible CSs.

Although it may well be true that there is indeed less inequality, it is not correct to conclude that there is less poverty or less underdevelopment, that is, if we accept the U.N. definition of 'least developed'. As defined by the U.N., 'least developed' represents a combination of extremely low per capita GNP (US$125 or less), a very high illiteracy rate (over 80 per cent among those 16 years and older) and little industrialisation (10 per cent share of GNP or less). This implies, among other things, low productivity, lack of economic diversity and few employment opportunities outside agriculture. Because Dependencia models like that expounded by Frank are unable to account for the poverty syndrome of the LLCs in the Third World, we must look for an alternative explanation.

AN ATTEMPT AT EXPLANATION

Might it be that the lack of economic growth and development in LLCs has to do with the fact that in the past they were, and today still are, the areas least affected by foreign capitalist influence, and, thus least integrated into the capitalist system? An affirmative answer to this question would turn Dependencia theory upside down, or would at least suggest that it needs to be amended. However, instead of jumping to this conclusion, we would be well-advised to return to the question of how to explain the near absence of development in most of the Third World's LLCs. In order to be able to arrive at an explanation, it seems imperative that we discover whether the LLCs share certain inherent disadvantages.

To begin with, LLCs tend to be located relatively far from the sea, even though some of them are not as far inland as are large parts of Sudan, Zaire, Algeria or Brazil. This distance factor forms a handicap due to the resulting high cost of transportation. For some commodities, shipping by air may be a solution, but for many it is not, so that production for export faces stiff competition from more favourably situated CSs. As a consequence, LLCs are frustrated in their industrialisation attempts. Expensive imported building materials, moreover, make construction of factories, bridges and power dams costly, thereby raising industrial overhead expenses and retarding infrastructural improvements. A poor physical infrastructure slows down industrial development and scares off potential foreign investors.

A more serious handicap than this distance factor, no doubt, is that LLCs do not possess (their own) ports since port locations have proven to be particularly well-suited for industrial development. In Africa as in other Third World regions, manufacturing is heavily concentrated in and near port cities, thanks to superior accessibility, cheap transportation and excellent opportunities for economies of agglomeration. This is especially true of industrial activities (e.g. assembly plants) which require imports of (bulky raw) materials or which produce for overseas markets. Simply the fact that ports constitute break-of-bulk points, makes them into favourable locations for the processing of raw materials. This explains not only why most oil refineries (plus associated petrochemical industries) are found near the coast, but also why agricultural raw materials originating from LLCs are often processed in neighbouring CSs rather than in the LLCs themselves. Clearly, then, CSs have important locational advantages, allowing greater profit margins and enabling them to attract foreign entrepreneurs or encouraging domestic entrepreneurs.

Considering the above, it is not surprising that CSs tend to be more highly industrialised than LLCs and that manufactured goods are more likely to move from CSs to LLCs than in the opposite direction, while raw materials tend to flow more from LLCs to CSs. For example, Selwyn (1975: 79) has pointed out that in the trade between Upper Volta and the Ivory Coast, 80 per cent of Upper Volta's exports to the Ivory Coast in 1969 consisted of live animals and animal products and a further 15 per cent of vegetable products. Ivory Coast's exports to Upper Volta were more diversified and included cement, wood and cork products, textiles, chemical products, transport products, food, drink and tobacco products and base metal products. This trade structure clearly illustrates the peripheral relation of Upper Volta to the Ivory Coast economy.

Moreover, such trade patterns pose a threat to the LLCs in the sense that they can cause them to become the victims of 'unequal exchange'. Stated differently, LLCs are more likely than CSs to have 'hinterland economies', supplying rather cheap (agricultural) raw materials to the somewhat more industrially developed CSs. And while the LLCs are less industrialised, they usually also have a lower per capita GNP, more limited tax revenues, poorer public services, lower literacy rates and smaller domestic markets. Furthermore, because of limited employment possibilities and low wages, many people migrate to the more dynamic neighbouring CSs in search of (better-paid) work. Thus, in their relations with the CSs, the LLCs not only perform a hinterland function but also serve as reservoirs of cheap labour.

In this light, we might characterise a Third World LLC as a 'double dependent periphery' (Blaikie 1981), occupying a subordinate and dependent position both with regard to one or more neighbouring CSs and to the world's capitalist core area.[4]

Possibly the greatest dilemma faced by the LLCs is their dependence on one or more neighbouring states for access to the sea. This dependence involves the use of foreign ports, of transportation links to those ports as well as of various services provided by the country (or countries) through which the coast must be reached. Without permission to use these facilities, a LLC cannot participate effectively in international trade. Because of the paucity of adequate routes to the sea in most of Asia, Africa and South America, it is obvious that the Third World's LLCs find themselves in a weak and vulnerable position vis-a-vis the CSs, something the latter can take advantage of. To obtain the required permission and to keep it once it has been obtained, it is essential that the LLC maintain friendly relations with its neighbour(s). A case in point is Malawi, which, mindful of its dependence on the ports of Nacala and Beira, has no choice but to be cordial to Mozambique. This situation is appropriate today under the Machel regime as it was when the Portuguese ruled Mozambique until 1975. Even so, there is always the possibility that relations inadvertently deteriorate, causing the CS to revoke its permission to allow 'free and innocent' passage. Such a decision amounts to closure of the common boundary. Short of closing the border, the CS can impose upon the LLC certain restrictions regarding the use of its transit links and port facilities, as when Tanzania banned heavy copper-trucks from Zambia from its highways. Such restrictions can be detrimental to the economy of the LLC, possibly resulting in inefficient use of (scarce) transportation means, partial unloading and reloading at the border causing delays and possible damage to merchandise (e.g. spoilage). Besides, the CS could raise substantially freight rates and other fees and tariffs. It is in this context that the barrier function of boundaries separating sovereign states manifests itself clearly.

263

Directly associated with a LLC's dependence on (rail)roads through and ports in a CS is its dependence on that same transit country for adequate maintenance of these facilities and related services. If proper care is neglected by the CS (or by an intervening transit state such as Uganda in the case of Rwanda's access to Mombasa, Kenya), the LLC will be the victim, causing its already weak competitive position in the world market to become even weaker. And if the CS is unwilling or unable to provide the special services needed by the LLC, e.g. refrigerated storage for perishable export commodities, or is incapable of preventing serious port congestion, the LLC will be adversely affected. What makes it so frustrating for the latter is that it cannot really do anything about these problems. In fact, it can do little more than try to maintain the best possible relations with the transit country. But this is often easier said than done, in particular when two countries adhere to conflicting political ideologies. All this adds an element of considerable uncertainty to the LLC's already precarious situation, above all the uncertainty that the CS may sooner or later take measures that are harmful to the LLC's economic interests.

Undoubtedly, one of the worst things that can happen to a LLC is that the CS (or an intervening transit state) closes the boundary, something that the dependent LLC must try to avoid at all costs, especially if it has access to only one outlet to the sea. The fact that the CS is aware of this only adds to the LLC's vulnerability. A closed boundary can lead to critical shortages of fertiliser, gasoline, food and other basic commodities that can force the LLC's economy to some to a halt.

Occasionally, the uncertainty is intensified or underlined by explicit reminders by the CS that the LLC is dependent on an open boundary. In Africa, such reminders have been issued repeatedly during the past two decades and they were usually meant to be poorly concealed threats. A single example will suffice. Newsweek (August 24 1981: 19) reported that

> '. . . Mugabe is unlikely to push South Africans too far. About 80 per cent of Zimbabwe's trade flows through South Africa; even its communications, including sensitive diplomatic traffic, are channelled through Cape Town and Pretoria. As one South African military man put it: "If we closed the South African–Zimbabwe border and bombed the tenuous rail link with Mozambique, Zimbabwe would be bottled up".'

To this may be added that every past border closure by a CS functions as an unpleasant reminder that a CS can put considerable political pressure on a land-locked neighbour, or blackmail it and take economic advantage of its dependent, exploitable condition. Understandably, borders are seldom closed by LLCs.

Another problem with which the LLCs are faced is that they can become the innocent victims of actions by a third party against the transit state on which they depend for access to the sea. In April 1979, Zimbabwe-Rhodesia blew up the Kazangula ferry between Botswana and Zambia in an effort to stop alleged shipments of arms to ZAPU in Zambia. As a result, Malawi lost its only road connection to South Africa and became, once again, completely dependent on routes through Mozambique. This was a serious handicap for Malawi because the ports of Nacala and Beira were congested, while at the same time, the Mozambican National Resistance Movement (MNRM) as well as Zimbabwe-Rhodesia were sabotaging roads, railroads, bridges and fuel depots in Mozambique. The consequence was that Malawi suffered an acute fuel shortage in the latter part of 1979 (Legum 1981: B697, B720-21, B744, B948).

In somewhat similar fashion, Rwanda and Burundi became victims of the war between Tanzania and Uganda in 1978 and early 1979, which all but crippled their trade. Burundi also complained that trains intended for its use had been taken over by Tanzania to service the needs of its military. Meanwhile, Rwanda reported that 130 trucks and at least 34 fuel tankers were held up in transit and that many were impounded by the forces of Idi Amin. Among the imports blocked along the way or in Mombasa were 1 million litres of petroleum products and 1,720 tons of foodstuffs. By March 1979, Rwanda had virtually run out of gasoline supplies, reason for the government to institute stringent rationing regulations (Legum 1981: B164, B285).

The African country most seriously affected by its unfavourable geopolitical position is undoubtedly Zambia. Over the years, it has suffered from closure of the Benguela Railway since 1975 as a result of the Angolan civil war; closure of the Rhodesia-Mozambique border (1976-80); the largely self-inflicted closure of the Zambia-Rhodesia border (1973-78 for the rail route and until 1979 for the road links); congestion of the ports of Dar-es-Salaam, Nacala, Beira and Maputo; labour strikes in the port of Lobito prior to 1975; insufficient storage at Dar-es-Salaam; destruction of the Kazangula ferry (1979-80); various attacks by Zimbabwe-Rhodesia (allegedly supported by South Africa) on Zambian roads, railroads and bridges, thereby cutting all of Zambia's main links with Mozambique and Tanzania (including the TanZam railroad); poor road conditions in Botswana, Mozambique and Tanzania; the liberation war in Mozambique prior to 1975; numerous acts of sabotage by MNRM in Mozambique since 1975; friction with Malawi; and differences of opinion with Nyerere and Machel following Zambia'a announcement in October 1978 that it would start using the rail route through Rhodesia to import desperately needed fertilisers and other goods, and export a growing stockpile of copper.

In its attacks on Zambian routes to the coast, Zimbabwe-Rhodesia made certain that Zambia's links with Zimbabwe-Rhodesia were left intact, a calculated attempt to make Zambia completely dependent on these outlets to the sea. And in the past few years, observers have become more convinced that South Africa has been making Zambia (as well as Zimbabwe) more dependent upon its transportation system and ports. By giving support to UNITA, it helps to keep the Benguela railway closed and by aiding MNRM, it obstructs the use of Mozambique's ports by Zambia (and Zimbabwe). Thus, Pretoria frustrates attempts by Zambia (and Zimbabwe) to become less dependent on South Africa. So far, it has been extremely successful and today, a very large proportion of Zambia's total foreign trade passes through South African ports. As long as this pattern does not change, South Africa will be able to put about as much pressure on Zambia as it can exert on the other land-locked countries of southern Africa: Lesotho, Swaziland, Botswana, Zimbabwe and Malawi. For the time being, all six countries will remain hostages of South Africa, a situation which enables Pretoria to pursue its policy of creating a 'constellation of southern African states'. And the existence of this constellation, in turn, reduces the chance that an oil embargo will be instituted against South Africa.

LIMITED DEVELOPMENT POSSIBILITIES

Glassner (1978: 19) has observed:

'Since most of the states across which developing land-locked countries must transit are themselves poor, they frequently find their ports and transport facilities inadequate for their own expanding needs, to say nothing of having a surplus capacity for the use of a land-locked neighbour. Delays in delivering transit traffic of land-locked states often result from . . . shortages of competent staff, inadequate and unreliable communications. . . .'

Cervanka (1973: 19) has come to a similar conclusion:

'. . . in the case of a land-locked country even a spectacular expansion of its economy would be meaningless if not accompanied by a comparable improvement in the communication links of its neighbours on which it depends'.

In fact, I would argue that substantial economic growth in a LLC is likely to result in increased dependence due to expanded shipments to and from overseas countries (see Reitsma 1980: 138). Be this as it may, many analysts (e.g. Szentes 1973; Hveem 1973; Pedersen and Leys 1973) agree that LLCs will find it difficult, if not impossible, to raise themselves above the development level of neighbouring CSs. This is true even when they are richly endowed with natural resources, possess a good physical infrastructure and are blessed with a favourable social and ethno-demographic structure. Switzerland, Austria and Luxembourg would not have been able to become the highly developed countries that they are had they not been surrounded by developed countries providing a dense and integrated network of high-quality infrastructural facilities. As mentioned earlier: no developed country is an island.

Conversely, the slow and difficult economic development of LLCs cannot have a beneficial long-term effect on the development of neighbouring CSs. Because the African continent has so many LLCs, it will be hard put to reach reasonably high levels of material well-being. Put differently, Africa's low level of development is in no small measure attributable to its high degree of political fragmentation or to the barrier effects of its multitude of international boundaries. And especially as long as relations between neighbouring states remain as bad as many of them have been since constitutional independence was attained, resulting in frequent border closures, Africa's excessive political fragmentation is likely to be a formidable obstacle to future economic growth. Further complicating this unhappy state of affairs is the circumstance that Africa's LLCs, like those on other continents, occur in clusters. The less developed its land-locked neighbours are, the more difficult it is for a given LLC to attain a respectable level of development. Much of interior Africa, therefore, can be expected to continue to be one of the world's least developed regions for many years to come.

CONCLUSIONS

It makes sense for development geographers to divide the countries of the Third World into two groups: coastal and land-locked. Because of their disadvantageous spatial situation, the latter have more limited possibilities for economic development than the former. This comparative disadvantage will continue so long as the continents on which they are located do not possess dense and integrated networks of transportation lines and a large number of modern, efficiently functioning ports, and above all, so long as LLCs and CSs do not improve their relations and cooperate more than they presently do.

Because the LLCs lag behind the CSs in economic development, they may well fall behind even further. Their doubly dependent predicament would seem to offer little hope that they will catch up in the foreseeable future. In the event they do experience considerable economic growth, the CSs may well try to benefit from this by looking for ways to extract part of the surplus. In addition, the fact that many of their young people, including some of the better-educated, have left for greener pastures elsewhere, further exacerbates the plight of the LLCs.

Considering the many disadvantages that LLCs are faced with as well as the development problems associated with excessive political fragmentation, one can hardly escape the conclusion that the creation of 'homelands' by South Africa augurs little good for the future of much of southern Africa. Not only will most of the homelands be land-locked, but several of them will consist of a number of small, separate, land-locked territories completely surrounded by areas reserved for whites. The result will be an abundance of boundaries which are bound to have negative effects on development and regional integration and cooperation.

Finally, it would appear that the distinction between (dependent) LLCs and (dominant) CSs offers possibilities for amending and refining existing core-periphery models and dependency theories by making them more geographically relevant and at the same time by bringing them more closely in line with 'the harsh facts' of reality.

FOOTNOTES

1. The number of these 'sub-imperialism studies' may well grow because it is becoming increasingly evident that intra-Third World (dependency) relations are important for understanding patterns of underdevelopment. In a recent study, Blaikie (1981: 244) observes that 'The "Indian connection" has been significant in the development of underdevelopment in Nepal.'
2. According to the most recent U.N. definition of 'least developed', there are 31 countries that qualify. No fewer than 15 of these are land-locked, i.e. Botswana, Burundi, Central African Republic, Chad, Lesotho, Malawi, Mali, Niger, Rwanda, Uganda and Upper Volta in Africa, and Afghanistan, Bhutan, Laos and Nepal in Asia.
3. Because of a dearth of statistical information on Swaziland, it is not clear how this country compares with the 31 poorest countries.
4. Incidentally, a country like Rwanda, whose one and only effective route to the sea passes through land-locked Uganda and coastal Kenya, may be said to form a 'triple dependent periphery'. To some degree, the same applies to Burundi, Chad, Zambia and Botswana.

REFERENCES

BLAIKIE, P., J. CAMERON & D. SEDDON [1981]
'Nepal: the crisis of regional planning in a double-dependent periphery', in Walter B. Stohr and D.R. Fraser Taylor, (eds.), Development From Above or Below? The Dialectics of Regional Planning in Developing Countries, (Wiley, Chichester), 231-58.

268

CERVANKA, Z. [1973]
 'The limitations imposed on African land-locked countries',
 in Z. Cervanka (ed.), Land-locked Countries of Africa,
 (Scandinavian Institute of African Studies, Uppsala),
 17-33.
FRANK, A.G. [1978]
 'Development of underdevelopment or underdevelopment of
 development in China', Modern China, 4, 341-50.
GLASSNER, M.I. [1978]
 'The Law of the Sea', Focus, 28(4), 1-24.
HVEEM, H. [1973]
 'Relationship of underdevelopment of African land- locked
 countries with the general problem of economic development',
 in Z. Cervanka, (ed.), Land-locked Countries, 278-87.
LEGUM, C. (ed.) [1981]
 Africa Contemporary Record, Volume 12, (Africana Publishing
 Company, New York).
PEDERSON, O.K. & R. LEYS [1973]
 'A Theoretical Approach to the Problems of Land-locked
 States', in Z. Cervanka, (ed.), Land-locked countries,
 288-92.
REITSMA, H.A. [1969]
 Crop and Livestock Differences on Opposite Sides of the
 United States-Canada Boundary, (Ph.D thesis, University
 of Wisconsin-Madison).
REITSMA, H.A. [1971]
 'Crop and livestock production in the vicinity of the
 United States-Canada border', Professional Geographer, 23,
 216-23.
REITSMA, H.A. [1972]
 'Areal differentiation along the United States-Canada
 border', Tijdschrift voor Economische en Sociale Geografie,
 63, 2-10.
REITSMA, H.A. [1980]
 'Africa's land-locked countries: a study of dependency
 relation', Tijdschrift voor Economische en Sociale Geografie,
 71, 130-41.
SELWYN, P. [1975]
 'Industrial development in peripheral small countries',
 in P. Selwyn (ed.), Development Policy in Small Countries,
 (Croom Helm, London), 77-104.
SZENTES, T. [1973]
 'The economic problems of land-locked countries', in Z.
 Cervanka (ed.), Land-locked countries, 273-77.

Chapter 21
IDENTIFYING REGIONAL ALLIANCES AND BLOCS IN THE UNITED NATIONS VOTING — SOME PRELIMINARY RESULTS OF VOTES AFFECTING ISRAEL

Stanley D. Brunn and Gerald L. Ingalls

INTRODUCTION

Within an ever-expanding community of nations numerous issues of a political nature arise each year. Compiling a list of such issues for all nation states would reveal some of greatest concern to the individual state itself, some to its neighbours and others to preservation and survival of all humankind. Clearly any effort to measure and assess an individual state's performance vis-a-vis internal and external political events is likely to become exceedingly difficult. Even if events of an internal nature are not considered, the task of compiling detailed annual inventories of how nations perceive and react to various international events and issues is staggering.

At issue here is not how the individual nation-state perceives itself but more how it perceives and is, in turn, perceived by the rest of the world. Such 'perceptions' can be obtained through an examination of roll-call voting in the U.N. General Assembly, the major international forum for the discussion of world issues over the past 35 years. An examination of an individual state's roll call voting record will provide some indication of how it views other states and international issues for which it is asked to express support or non-support. Conversely, examination of roll call votes will provide some indication of how other nation-states view the individual state in question.

This chapter has five objectives:
1) to introduce the topic of U.N. voting in the General Assembly and suggest it as a profitable and timely topic for inquiry in political geography.
2) to provide a literature review on the state of such analyses;
3) to discuss the data sets available for conducting geographical and temporal analyses of U.N. voting;
4) to demonstrate the utility of U.N. voting records by using some results of Israel's voting record in the Assembly and those of other states voting on issues important to Israel;

5) to suggest a series of topics meriting subsequent research.

LITERATURE SURVEY

The only studies by geographers on roll-call votes in the U.N. General Assembly have been by Abate (1976) and Abate and Brunn (1977) who investigated whether the voting results of African members in the U.N. General Assembly were directly related to the amounts of U.S. and Soviet assistance. They concluded that often foreign aid levels were inversely related to the votes, especially of former colonies.

Several major themes have been developed on U.N. voting during the past two decades. One of the dominant foci is the study of blocs within the U.N. Several authors have used analyses of roll-calls to identify the existence and persistence of blocs (Alker and Russett 1960; 1965; Bailey 1961; Ball 1951; Goodwin 1976; Hovet 1960; Manno 1966; Newcombe et al. 1970; Russett 1966; 1967; Singer 1963; Vincent 1965; 1969; 1970; 1971; 1972; Weintraub 1977). Most of these studies have attempted to uncover blocs using the voting record of all U.N. members.

Foreign policy and aid questions, insofar as they affect voting behaviour of U.N. members, have been another major theme. Rai (1972) and Wittkopf (1973) address these questions on a general level while Alpert and Bernstein (1974) and Bernstein and Alpert (1971) address the strategies of the U.S. to thwart China's admission into the U.N. Keating and Keenleyside (1980) examined Canadian votes in the General Assembly on major colonial and racial issues.

Other studies have addressed major issues facing the international community. Human rights and anti-colonial stances have been major issues uniting many newly independent states. U.N. votes on these issues have been studied by Hurwitz (1976), Rowe (1964; 1970) and Shay (1968). China's voting behaviour, which is more favourable to Third World members than the major superpowers, has been examined by Chai (1979). Ministates, which have also mixed voting patterns, have been studied by Harbert (1976). Cold War issues in the U.N. have been examined by Gareau (1968-1969; 1970) in detailed studies. The various votes on U.N. seabed disputes have been studied by Friedman (1970). Clark et al. (1971) attempt to define some general attributes of states' support of the U.N.

A final significant area of U.N. voting research has focused on how to organise, rank and interpret the votes of individual states. Factor analyses, cluster analyses, canonical analyses, Guttman scaling techniques and other indices have been utilised to identify blocs and relationships between and among U.N. members. Some authors including Alker (1964), Alker and Russett (1960; 1965), Russett (1966, 1967), Newcombe et al. (1970) and Vincent (1971, 1972) have assigned weights to votes: no=1, abstain=2, yes=3. Others, including Lijphart (1963) do not consider absences or abstentions in defining indices. Studies using multivariate statistical techniques identify salient groupings of votes or states that formed blocs. Both R-mode and Q-mode analyses exist in the literature.

Data Sets

The primary source for votes in the U.N. is the annual Index to Proceedings of the General Assembly (United Nations). Another major source is the data published by the Canadian Peace Research Institute in Dundas, Ontario. We have compiled data from the first two of the six volumes published by this organisation (Schopen et al. 1975; Newcombe and Wert (1979) and the Supplement (Newcombe and Mahoney 1981).

The first volume contains data on U.N. roll call votes from January 1946 to December 1973. Data are organised chronologically by yearly assembly session and the volume is divided into two major parts: (a) a description of each resolution and (b) the voting record of each nation. The second volume contains single pairwise Pearson product number coefficients in which the vote of every U.N. member on every issue or resolution is compared to the votes of all other members on the same issues. The correlation matrices provide a wealth of data for political geography investigations such as studies of friends, adversaries, neighbours, political and economic alignments and political and ideological turnarounds.

ANALYSES

Selected Individual U.N. Votes

The volume Nations on Record offers political geographers a rich source of well-organised data on the voting behaviour of individual states to test the existence and formation of political regions, blocs or supra-national organisations. To demonstrate the utility of this data source, we have selected the results of five specific roll call votes. Four of these votes focus on Israel or Palestinian rights while the fifth concerns China.

The five individual Assembly roll call votes examined are those resolutions which called for the:
1)   Admission of Israel to the U.N. in 1949,
2)   Condemnation of Israel for aggression and the U.S.A. and the U.K. for assistance and direct participation in 1967,
3)   Admission of the People's Republic of China and the expulsion of Taiwan in 1971,
4)   Full respect for Palestinian self-determination as an indisputable element in Middle East peace, 1971, and

5)   Condemnation of the 'unholy alliance' between Portuguese colonialism, South African racism, Zionism and Israeli imperialism, 1973.
In each of these resolutions the voting was significantly split among yes and no votes and abstentions.

Analysis of the voting patterns on these particular resolutions offers, for example, the reactions of various regional and ideological groups on resolutions pertaining to Israel. Previous analyses of U.N. voting have revealed that the cohesion of regional or ideological associations can vary by issue. Regional groups in particular, serve mainly as a means of consultation and communication and do not necessarily dictate unified voting behaviour (Keohane 1966). Bailey (1961: 29) suggested that the only legitimate bloc within the U.N. was the Soviet bloc consisting of the Communist member states of the Warsaw Pact and the Council for Mutual Economic Assistance together with the Byelorussian S.S.R. and the Ukrainian S.S.R. Despite the popular conception of bloc voting on matters pertaining to Israel and Palestine by Arab, Moslem and Afro-Asian members, we would do well to heed Bailey's recognition of the impermanence of U.N. groups or blocs. By mapping and analysing the votes on these four resolutions pertaining to Israel and that on the controversy surrounding the seating of China, we can compare the degree of cohesion within such regional associations as Western Europe, Latin America, the Afro-Asian community, the Islamic community and the Soviet bloc.

The vote on the admisson of Israel came in May 1949. The resolution was accepted with 12 no and 37 yes votes with 9 abstentions. As can be seen in Figure 21.1, the pattern of voting fits closely with the conception of bloc voting in the U.N. All of the 12 nations voting no were clustered within South and Southwest Asia. The solidarity of this group on this resolution was, in all probability, a lingering aftermath of the conflict surrounding the birth of Israel, the 1948-49 war and the affinity of other Moslem states with Israel's neighbours. It is worth noting that in 1950, this group (which is at present the largest single organised group at the U.N.), was formally recognised. Of the 12 states which voted no on this resolution, only Ethiopia was not among the original founding member states of the Afro-Asian group. And as Bailey (1961: 34) points out, the most cohesive cluster within this group were the Arab states, who had been in close consultation on various Middle Eastern questions

In examining the results of the voting on this resolution, we can foresee the formation of an important regional and cultural association within the U.N. Other groups existed before and others have followed since. As more states became independent and gained U.N. membership, regional or ideological groups, including the Afro-Asian group have served to crystallise a regional identity within the General Assembly; such groups have had significant consequences on U.N. voting patterns. However, as is apparent in examining the other resolutions, the cohesiveness of these regional groups varies across time and by issue. The 1967 resolution to condemn Israel for aggression offers some evidence of this effect.

The specific resolution examined was but one of fifteen roll call votes taken on 4 July 1967 during an emergency session. Eight of these votes were on resolutions or amendments condemning Israel for aggression. All eight were rejected. The wording of this resolution - condemning Israeli aggression, blaming the U.S. and the U.K. for aiding, inciting and participating in the war, declaring Israel fully responsible and demanding full reparations - certainly played a role in the strength of the opposition. On this particular vote the Afro-Asian vote was quite fragmented. Indeed, there was little cohesion even among the Arab Islamic nations. Iran, Turkey, Libya, Tunisia and Pakistan chose not to support this resolution. Only Israel's immediate neighbours, the Soviet bloc, Cuba, Cambodia, Algeria and Mauritania endorsed this strongly-worded resolution. Western Europe and the entire western hemisphere (with the exception of France, which abstained) voted to reject the resolution.

Two votes taken in 1971 offer further evidence of the rather transitory nature of U.N. voting blocs or groups on critical issues. In October, the General Assembly voted 76 to 35 with 17 abstentions to admit the People's Republic of China and expel Taiwan. Later in the year, the General Assmbly, by a vote of 53 to 23 with 43 abstentions, passed a resolution declaring that respect for Palestinian rights was an indispensable element in establishing peace in the Middle East.

On the China representation vote, several key regional or ideological groups voted in concert. Western Europe, the Soviet bloc, the Islamic states and Israel voted to accept the resolution. Of these three groups, only four abstentions and one no vote (Saudi Arabia) broke the regional solidarity. On the other hand, the African states and the member states in the Western hemisphere demonstrated little unity.

On the Palestinian self-determination issue, virtually every major voting group witnessed significant splits. Only the Soviet bloc remained intact. Western Europe, in particular, showed significant splits with 5 no and 2 yes votes and 8 abstentions. Similar divisions occurred within the Latin American and African groups. Israel voted no. (Figure 21.2).

In December 1973 an amendment to a previously considered resolution was introduced calling for a condemnation of 'the unholy alliance between Portuguese colonialism, South African racism, Zionism and Israeli imperialism'. The amendment was accepted with 32 no votes, 63 yes votes and 27 abstentions. This particular vote was taken on an amendment to a previous resolution condemning South African apartheid, in which only one state, South Africa, voted no. The pattern of voting on this amendment (Figure 21.3) offers evidence that on some issues, regional groups can coalesce quickly. Once again, we see evidence of fragmented voting among Latin American states. Conversely,

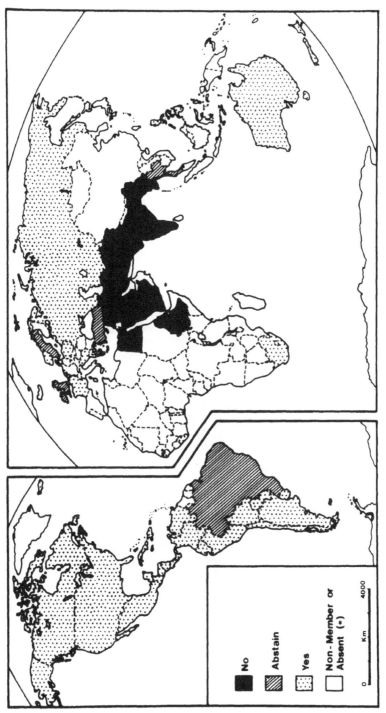

Figure 21.1: Admission of Israel to United Nations, 1949

No
Abstain
Yes
Non - Member or
Absent (*)

Km
0   4000

275

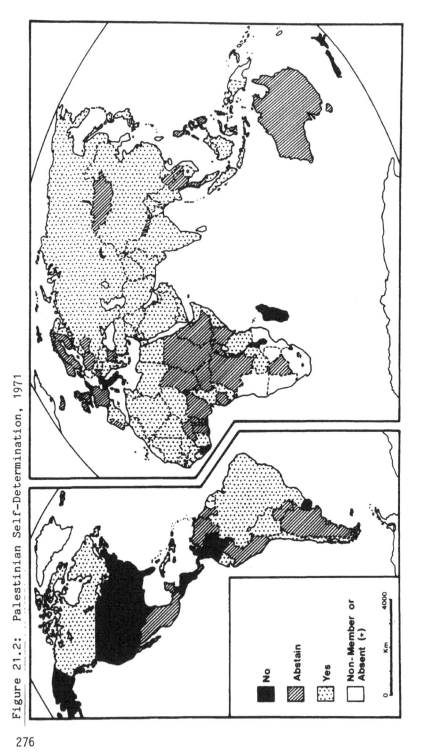

Figure 21.2: Palestinian Self-Determination, 1971

Figure 21.3: Portugal, South Africa and Israel, 1973

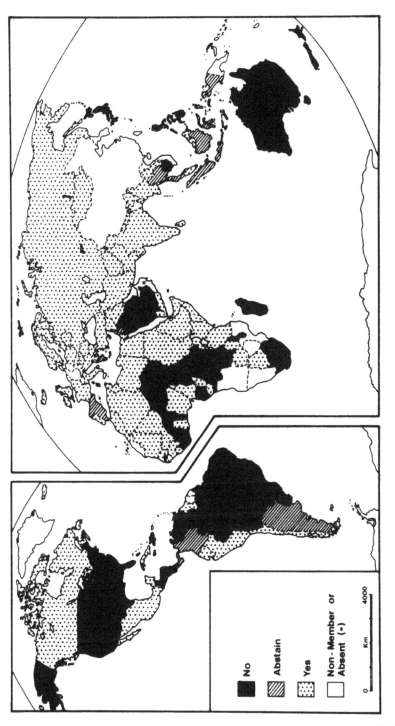

No

Abstain

Yes

Non-Member or
Absent (•)

0   Km   4000

Western European states are unanimous in their rejection of this amendment. Equal, if opposite, solidarity is apparent within the Afro-Asian group. While 17 states did abstain, in all of Africa and Asia, only two states – Israel and South Africa – voted to reject this amendment.

At least two conclusions may be drawn from our analysis of the pattern of voting on these five roll call votes. First, there appears to be rather clear and consistent evidence that on these types of issues regional blocs are seldom clearly defined. The Soviet bloc is perhaps the only exception to this. On matters pertaining to Israel, where resolutions often prompt considerable divisiveness, our maps and analyses reveal that division most often cuts across regional groups. It also appears likely that the analysis of single individual votes may produce quite misleading impressions. Consider, for example, the conclusion one might draw from the votes on Palestinian rights and the condemnation of Israel, Portugal and South Africa. In one instance regional groupings appear important; in another, they seem irrelevant.

Clearly, there are pitfalls involved in the use of single roll call votes to generalise about the behaviour of particular states or groups of states on certain international issues. There is no certainty that the vote taken on any given issue is not aberrant behaviour. If the objective of such analysis is to assess consistency of dynamism, then a longer term data base is required.

PATTERNS OF CORRELATION COEFFICIENTS FOR SELECTED SESSIONS

The Pearson coefficients between Israel and all U.N. members are grouped into five classes and mapped to discern whether any meaningful patterns emerge. The four sessions examined are Numbers 4 (1949), 23 (1968), 29 (1974) and 32 (1977). The sessions were selected to discover those states that agreed with Israel or vice-versa in its first year in the U.N., those states that voted as it did following the 1967 and 1973 Middle East conflicts and the most recent year the pairwise correlations are available (1977).

The analysis reveals some consistency in voting patterns and also a surprising amount of dynamism. The map for the 1949 session showed the pattern was not consistent among members from Western Europe, Latin America or Asia. Some of the highest correlations were with states in North America, Central America and southern and eastern Asia. Eastern European countries and the Soviet Union had negative correlations as did Australia, Ethiopia, France and Belgium. Guatemala, Chile, Canada, Iceland, Turkey and Thailand were among those with the highest correlations.

The 1968 votes displayed some regionality. Most states in Western Europe, North, Central and South America, as well as

Japan, Australia and New Zealand voted similarly to Israel or vice-versa. The largest bloc of states voting in opposition to those resolutions which Israel supported were in Eastern Europe, U.S.S.R. and pro-Russian governments, such as Mongolia, Cuba, Mali and Guinea. The Afro-Asian 'bloc' and that of Middle East and North African Arab states displayed some diversity. Indonesia, Singapore, Chad, Morocco, Tunisia and Saudi Arabia had positive correlations with Israel while Iraq, Sudan, Syria and Yemen had correlations greater than -0.20. A large number of newly independent states and new U.N. members had correlations near zero. Cameroon, Ghana, Upper Volta, India, Kenya, Zambia and Uganda are examples. A major feature of the 1968 voting was the lack of uniformity in votes among the Less Developed Countries.

The voting patterns from the votes in 1974 displayed much of the consistency and diversity shown in 1968. Members having votes most similar to Israel were in North, Central and South America was well as Western Europe, Japan, Australia and New Zealand. Canada, U.S., Netherlands and Denmark had the highest correlations. As in 1968, most of the highest inverse correlations were among Eastern European members, the Soviet Union and pro-Soviet satellites. The members with the highest negative correlations were India, Algeria, Burundi, Albania, Iraq and South Yemen. Examples of African and Asian states could be found with votes slightly positive (Philippines, Malaysia, Lesotho, Iran, Swaziland and Morocco) and highly negative (Afghanistan, China, Libya and Syria). Among Middle East countries there was some diversity although most had negative correlations.

With each successive session during the 1960s and 1970s the emergence and consistency of blocs becomes easier as more newly independent states were added to the U.N. membership. An examination of the votes for 1977 reveals some of the consistency apparent in 1968 and 1974. The U.S., Canada, Australia and selected Western European countries continued to have voting behaviours most similar to Israel. Sweden, Japan, New Zealand and many Central and South American states that had high correlations in the previous years now were included in the second highest category. Most African and Asian states were in those categories slightly above and below zero; this included former British and French colonies that were both large and small, rich and poor. Middle Eastern and North African states also had correlations near zero. The largest bloc voting opposite to Israel was the Soviet bloc of East European states, U.S.S.R., and several pro-Soviet states. The individual members voting most unlike Israel were Ukraine, Byelorussia, Poland, Bulgaria, Hungary, Czechoslovakia, East Germany and the Soviet Union itself; all had exactly the same correlations of -0.37.

FUTURE RESEARCH DIRECTIONS

The breadth and depth of previous research would seem to offer a number of possibilities for research by political geographers. We will identify four areas that seem promising. First, it would seem useful to examine the evolution and persistence of blocs over a one, two or three decade time frame. We need to know if U.N. voting patterns are a meaningful way to identify blocs. It may be that for some regions, military and economic ties are the factors which bond states together in their U.N. voting patterns. Conversely, such ties may not be reflected in U.N. votes. In this same area, we also need to identify what constitutes a bloc. Is it contiguity, similar histories or economic union? If we can identify blocs, it would also be worth finding out exactly what are the votes and issues that lead to their evolution and persistence.

A second topic worthy of inquiry involves the identification of votes where there was most and least agreement in a series of U.N. sessions and mapping the performance of member states on these votes. Are the types of resolutions which promoted the highest level of agreement in 1975 the same as in 1965 or 1955? Are the states that voted most like the majority in one year the same that voted most like the majority five, ten or fifteen years earlier?

A third research focus is on the correlation profiles of U.N. members. We have seen with the examples above the stability and fluctuations of selected states vis-a-vis Israel. How would the profiles look if we examined the profile of all African or Asian states with Israel or with Japan or with the Soviet Union? Would Subsaharan African countries with British influence have similar profiles? Do newly independent states have different profiles with major foreign aid donors than those that achieved independence in the 1960s? Do ministates have similar profiles with the U.S., U.S.S.R., West Germany or France? Are there similarities in the fluctuations of Central or South American countries vis-a-vis a major superpower?

A fourth and final topic is one that has been addressed to some extent and that is the relationships between foreign aid and U.N. voting. As noted above, several studies have been completed on aid from U.S., U.S.S.R., Israel and the Arab states to Africa. The results of these studies have shown that often an inverse relationship exists between aid and voting. We need to know more about the direction and strength of such relationships among Less Developed Countries in Africa, Latin America and Asia, not only in a few years, but for a couple of decades.

Further research into U.N. voting behaviour by political geographers will help us understand the variations that exist in the policies of individual states and the political machinery of the U.N.

REFERENCES

ABATE, Y. [1976]
  Foreign Aid, U.N. Voting Behavior and Alliances: The Case
  of Africa, the U.S. and the U.S.S.R., Unpublished Ph.D
  thesis, (Department of Geography, Michigan State University,
  East Lansing).
ABATE, Y. and S.D. BRUNN [1977]
  'The emergence of African voting blocs and alliances in
  the United Nations, 1961-1970', Professional Geographer,
  29, 338-46.
ALKER, H.R. Jr. [1965]
  'Dimensions of conflict in the General Assembly', American
  Political Science Review, 58, 642-57.
ALKER, H.R. Jr. and B.M. RUSSETT [1960]
  Bloc Politics in the UN, (Harvard University Press,
  Cambridge).
ALKER, H.R. Jr. and B.M. RUSSETT [1965]
  World Politics in the General Assembly, (Yale University
  Press, New Haven).
ALPERT, E.J. and S.J. BERNSTEIN [1974]
  'International bargaining and political coalition: United
  States foreign aid and China's admission to the United
  Nations', Western Political Quarterly, 27, 314-27.
BAILEY, S.D. [1961]
  The General Assembly of the United Nations: A Study of
  Procedure and Practice, (Praeger, New York).
BALL, M.M. [1951]
  'Bloc voting in the General Assembly', International
  Organization, 5, 3-31.
BERNSTEIN, S.J. and E.J. ALPERT [1971]
  'Foreign aid and voting behavior in the United Nations:
  the admission of Communist China', Orbis, 15, 963-77.
CHAI, T.R. [1979]
  'Chinese policy toward the Third World and the superpowers
  in the U.N. General Assmebly, 1971-1977: a voting analysis',
  International Organization, 33, 391-403.
CLARK, J.F. et al. [1971]
  'National attitudes associated with dimensions of support
  for the United Nations', International Organizaton, 25,
  1-25.
FRIEDHEIM, R.L. and J.B. KADANE [1970]
  'Quantitative content analysis of the United Nations seabed
  debate', International Orgaization, 24.
GAREAU, F.H. [1968-1969]
  The Cold War 1947-1967: A Quantitative Study, (University
  of Denver Social Science Foundation and Graduate School
  of International Affairs, Monograph Series in World Affairs,
  6, #1, Denver).

GAREAU, F.H. [1970]
'Cold War cleavages seen from the United Nations General Assembly', Journal of Politics, 32, 929-68.
GOODWIN, G. [1976]
'The expanding United Nations': 1 - Voting Patterns', International Affairs, 36, 174-87.
HARBERT, J.R. [1976]
'The behavior of ministates in the United Nations, 1971-1972', International Organization, 30, 109-27.
HOVET, T. [1960]
Bloc Politics in the United Nations, (Harvard University Press, Cambridge).
HURWITZ, L. [1976]
'EEC and decolonization - voting behavior of the Nine in the U.N. General Assembly', Political Studies, 24, 435-37.
KEATING, T.F. and T.A. KEENLEYSIDE [1980]
'United Nations: voting patterns as a measure of foreign policy independence', International Perspectives, May/June, 21-26.
KEOHANE, R.O. [1966]
'Political influence in the General Assembly', International Conciliation, 557, 1-64.
LIJPHART, A. [1963]
'The analysis of bloc voting in the General Assembly', American Political Science Review, 57, 902-17.
MANNO, C.S. [1966]
'Majority decisions and minority responses in the U.N. General Assembly', Journal of Conflict Resolution, 10, 1-20.
NEWCOMBE, H. et al. [1970]
'United Nations voting patterns', International Organization, 24, 100-221.
NEWCOMBE, H. and J. WERT [1979]
The Affinities of Nations: Tables of Pearson Correlation Coefficients of U.N. General Assembly Roll-Call Votes (1946-1973), (The Peace Research Institute, Dundas).
NEWCOMBE, H. and T. MAHONEY [1981]
The Affinities of Nations: Tables of Pearson Correlation Coefficients of U.N. General Assembly Roll-Call Votes (1974-1977) Supplement, (The Peace Research Institute, Dundas).
RAI, K.B. [1972]
'Foreign policy and voting in the U.N. General Assembly', International Organization, 26, 589-94.
ROWE, E.T. [1964]
'The emerging anticolonial consensus in the United Nations', Journal of Conflict Resolution, 7, 209-30.

ROWE, E.T. [1970]
'Human rights issues in the United Nations', Journal of
Conflict Resolution, 14, 425-40.
RUSSETT, B.M. [1966]
'Discovering voting groups in the United Nations', American
Political Science Review, 60, 327-39.
RUSSETT, B.M. [1967]
International Regions and the International System, (Rand
McNally, Chicago).
SCHOPEN, L. et al. [1975]
Nations on Record: United Nations General Assembly Roll-
Call Votes (1946-1973), (Canadian Research Institute,
Oakville-Dundas).
SHAY, T.C. [1968]
'Non-alignment Si, neutralism No', Review of Politics, 30,
228-45.
SINGER, M.R. and B. SENSING III [1963]
'Elections within the United Nations', International
Organization, 17, 901-25.
VINCENT, J.E. [1965]
The Caucusing Groups at the United Nations: An Examination
of Their Attitudes Towards the Organization, (Oklahoma
State University Press, Stillwater).
VINCENT, J.E. [1969a]
'An analysis of attitude patterns at the United Nations',
Quarterly Journal of the Florida Academy of Sciences, 32,
185-209.
VINCENT, J.E. [1969b]
'The convergence of voting and voting patterns at the
U.N.', Journal of Politics, 31, 952-83.
VINCENT, J.E. [1970]
'An analysis of caucusing group activity in the United
Nations', Journal of Peace Research, 2, 133-50.
VINCENT, J.E. [1971]
'Predicting voting patterns in the General Assembly',
American Political Science Review, 65, 471-98.
VINCENT, J.E. [1972]
'An application of attribute theory to General Assembly
voting patterns and some implications', International
Organization, 26, 551-82.
WEINTRAUB, S. [1977]
'How the U.N. votes on economic issues', International
Affairs, 53, 188-201.
WITTKOPF, E. [1973]
'Foreign aid and U.N. votes', American Political Science
Review, 67, 868-88.

Chapter 22
A GEOPOLITICAL MODEL FOR INTERNATIONAL BEHAVIOUR

Robert W. McColl

Geopolitics is a term that often evokes strong emotions and misunderstanding. The problem lies not with the term itself, but rather its historic association with Nazi Germany and the belief that somehow geopolitics was applied as an aid to world domination and to justify inhuman acts. However, in recent years, there has been a reintroduction of the term in pronouncements from the United States Government. Geopolitics is a perfectly legitimate and useful word with explicit meaning and simply refers to geographic factors that lie behind political decisions.

In contrast, Political Geography refers to political decisions that modify geographic patterns such as boundaries, forced migrations, or the distribution of food. Naturally, the two terms are closely related and often interwoven. A geographic fact such as the presence of a mineral deposit may generate a political decision (Geopolitics); this decision may create a new political decision, ad infinitum (McColl 1966: 143-46). The careful use of the terms Geopolitics and Political Geography thus can be used to clearly indicate general motivation, the base of political action or the nature of our analysis.

POLITICAL PERCEPTIONS AND GEOPOLITICAL REALITIES

In the past, geopolitical (strategic) models emphasised the medium through which power could be applied. The policy and geographic impact of models based upon perceptions of Sea Power, Land Power and Air Power resulted in such political features as the Panama and Suez Canals, the policy of 'Containment' and the creation of the Strategic Air Command. Geographically, the models were valid, but their political application depended upon the ability of a nation to remain secure in its homeland. Only so long as there was national security could political force and power be extended outward until it met other nations with equal or greater power (Jones 1955: 492-508).

Given current military technology, the application and perceptions of geopolitical models no longer can be limited to traditional models of sea power, land power or air power. Herz has demonstrated that the territorial security of states is now gone (Herz 1957: 473-93). The ability to launch missiles from ships or home bases to virtually any location on the earth, with or without a follow-up by ground troops, has marked the demise of the territorial state as a function of national defence. Today it is necessary to view the geopolitics and strategies of international politics in terms of the actual actions and reactions (behaviour) of nations rather than rely on dated models that today are solely of tactical value.

To prevent accidental catastrophe, to control the response of nations or groups to perceived threat, it is essential to identify and classify what can be viewed as the fight, critical and flight zones of nations, especially for those nations capable of initiating global war.

While the integrity of national territory may no longer be valid as a basis for national defence, this does not mean that nations have ceased to behave territorially. Associated with the defence of every nation, especially for the superpowers who are technologically the most capable of ignoring distances and of total global destruction, there are zones analogous to the fight, critical and flight zones identified by animal behaviouralists and others (Hediger 1955; Hall 1966; Eibel-Eibesfeldt 1975; Fox 1974).

BEHAVIOURAL MODEL OF GEOPOLITICS

Ignoring the multitude of political and psychological factors that can precipitate acts of aggression, let us focus on how such behaviour may be triggered by geographic factors. A major trigger of international conflict is the conscious, or accidental, 'invasion' of territories that another state or political unit considers essential to its survival. Locating these behavioural zones provides a geographical explanation and/or prediction for political behaviour, including the intensity of response to actual or perceived aggression.

It is possible to propose the following postulates for a behavioural model of Geopolitics:
1.   All states follow the prime directive of guaranteeing national security or survival of the state. With current technology (missiles and satellites), such defence may now be affected by actions far distant from the state's actual boundaries. Regardless of where a threat may occur, a primary motive of international politics is to make certain of national survival.
2.   States that must import raw materials critical to their industrial, military, or social needs (such as oil, minerals and food) must defend the locations and access routes to such materials

just as they would any other territory viewed as critical to
national survival. In such geographic areas, aggressive activities
or actions that might prevent access must be viewed as direct
threats to the state itself. It is imperative to guarantee no
interruption or enemy control of such areas.
3. Major (global) powers often engage each other in political
confrontations, or 'shoving matches', best described as a 'pecking
order' contest. This 'sabre rattling' and confrontation seldom
involves any direct threat to the actual existence of the state
and rarely leads to serious consequences, such as direct war
between the contending parties. This type of conflict most often
occurs in areas some distance from and not critical to the
survival of either party. Examples would include U.S. - U.S.S.R.
confrontations in Greece.
4. Most aggression is triggered by actual or perceived threats
in two distinct geographic behavioural areas: fight zones and
critical zones. Only beyond the areas of threat to survival,
in the flight zones, are emotions sufficiently controlled or low
enough to permit a sublimation of the survival motives and to
allow either or both parties to break off conflict without any
consequence (actual or perceived) to the survival of their
state.

It is in the context of these postulates that the locations
of fight, critical and flight zones of national behaviour can
be identified and mapped.

SELECTED EXAMPLES OF FIGHT, CRITICAL AND FLIGHT ZONES

Fight Zones are locations where the presence of an enemy
or any uninvited presence would be viewed as actual invasion and
a threat to the existence of the state. Threats or aggression
in such locations evoke the greatest emotional and nationalistic
response and behaviour. Any action or behaviour is justified
in preventing an enemy from actually entering or occupying such
areas. With modern technology, such a zone need not be contiguous
to the territory of the state itself. Perhaps one of the most
dramatic examples of the unintended consequences of a transgression
into a fight zone was the 1962 Cuban missile crisis between the
U.S. and the U.S.S.R. The Caribbean had long been considered an
American sphere of influence or protective buffer. The presence
of a foreign power has only occasionally been tolerated and only
if its actions were viewed as those of a friend and not a threat,
e.g. United Kingdom in Belize. The presence of an aggressive
enemy or its military equipment in a client state would not be
tolerated. The Cuban missile crisis clearly demonstrated what
occurs when such a 'fight' zone is invaded, even accidentally.
The U.S. was clearly prepared to go to war if the Russian missiles
were not removed. It perceived its very existence, especially
the security of its economic and political core area - the eastern
seaboard - as threatened. There was a unified national response

286

that could have and would have accepted _any_ military response. The U.S. did not threaten Cuba to resolve the problem. It addressed the U.S.S.R. directly. The U.S.S.R. wisely recognised the seriousness of U.S. emotions and removed the missiles.

Similarly, during the Korean conflict, the People's Republic of China stated it would not passively accept American or U.N. troops north of the 38th parallel. The Chinese could not, and would not, tolerate an aggressor and self-proclaimed enemy close to a territory considered vital to its very existence (industrial Manchuria). When U.N. forces crossed the parallel, China entered the conflict as it had predicted it would.

Critical Zones historically were contiguous to national boundaries. When such areas were entered (invaded) by a potential aggressor, they evoked a sense of extreme wariness, a need to decide whether to fight or flee (retreat). The least error in judgment could trigger an emotional shift to a fight mode, an all out attack in defence of national survival. Such locations were indeed critical and often precipitated what was viewed as defensive conflict. In fact, the critical zone could be viewed as divided into two areas. One is most prone to fight response. The other would be associated more with areas more prone to negotiation or flight.

The contemporary behaviour of Israel in the Golan Heights and in Lebanon can be viewed in this context. The location as well as the proclaimed intentions of the former occupants of these territories made each location crucial to Israel's physical survival. Thus, Israel's control or 'neutralisation' was viewed as essential, but withdrawal under proper safeguards would not be impossible.

Critical zones are the most difficult to define and locate in geopolitical terms. Often they are similar to fight zones with any invasion precipitating an immediate and emotional response of self-defence. The major distinction from the fight zone is behavioural; there is a vacillation between aggression and a willingness to negotiate. A fight zone brooks no negotiation. It must be controlled for survival.

Under one set of conditions (e.g., primitive technology), an area may be merely a critical zone; under another (long-range missiles), the same area becomes a fight zone. This is what makes the line between fight and critical zones the most dangerous and difficult to deal with in terms of expected reactions. They are highly volatile and seem to shift between a willingness to negotiate and an all out national commitment to war.

For example, the critical zone along the sea boundary of the state customarily was the distance a ship could fire a cannonball - about three miles. This eventually expanded to twelve miles. Today, modern missile frigates and submarine-launched missiles can reach far beyond this distance. The current shift to a 200-mile limit, while ostensibly for economic reasons, clearly gives at least some warning time should an attack be launched by sea. It is an expansion of the fight-critical zone.

Flight Zones occur in locations sufficiently unimportant to conflicting states that survival emotions are very low. National prestige may be at stake but seldom is national survival endangered. There is no value in defending such areas to the death.

These are perhaps the most intriguing zones. Often they appear to involve the same emotions and potential for conflict as other geographic areas, but distance and lack of vital importance reduces actual commitment to their defence. This was President de Gaulle's assessment of the position of France in NATO. He was certain that the U.S. would not risk its survival for that of France in a nuclear war. It is a view common in the rest of NATO today. It also is a perception that American allies, such as Israel, try to offset by 'demonstrating' their value to the 'survival' of America - to create an emotional commitment.

In flight zones, enemies may challenge, call names, send troops, even engage in battle, but seldom does the escapade or scenario lead to actual war - although war may be invoked as a threat or bluff in the jockeying for position and prestige.

Perhaps the most recent examples would be the conflicts in Vietnam and Angola. In neither instance did either the U.S. or U.S.S.R. view the conflict as a threat to the other's home territory. Each side was directly engaged in conflict, but they were conflicts that posed no actual threat to national survival - although many died. Contrast these conflicts with events in Poland, Afghanistan or El Salvador, where much greater stakes are perceived.

In order to identify the attitudes associated with specific locations at different points in time, a number of methods can be used. First, we may use news content analysis. Nationwide public pronouncements in the printed, radio and television media should be reviewed for stories that deal with other nations or geographic areas. The quantity as well as quality of such news can provide a good first approximation of which areas of the world are of greatest concern. Items to consider include:

1. Column inches   The amount of space (quantity) devoted to various geographic locations provides a rough measure of interest. Quantity of reporting either indicates, or promotes, an awareness of the location and any problem associated with that area.

2. Tone   The tone (quality) of the news may indicate emotion associated with an event and area. For example, when there is a cessation of discussion following a prolonged period of intense and highly charged emotional outbursts, it is likely that talk is now over and action (fight) will now occur. This was especially the case preceding Israeli actions in Lebanon and Egypt, and Russian actions in Hungary and Afghanistan. When absolute statements are cast in moral terms or when there is a dehumanisation of the 'enemy' and a focus on details of actions viewed as a threat to national survival, or when the situation is termed a

conflict versus a controversy, conflict is probable, often imminent.

When news reports cover both sides of the issue and there is discussion of the merits or there is editorial discussion of the need to avoid unnecessary conflict, then there is unlikely to be a conflict. Recent European and American news coverage of the behaviour of the U.S. and the U.S.S.R. clearly indicates no desire by the Europeans or Americans to be any part of a conflict in Europe.

3. Maps   The use of maps often indicates a growing emotional fixation on some geographic location. More importantly, the amount of detail and the nature of the maps often either reflects or creates emotional intensity. For example, when an issue is indicated by a general map of the entire region or country (Middle East, Iraq-Iran, South East Asia or Central America), then national opinion is being focused both on the area as well as on the events in the area, but not with such intensity that conflict would be likely. Still, maps set the stage for an escalation of national involvement. Once maps of an individual country appear, when they contain the symbols and indications of conflict (such as burst symbols, military units and deployment, missile locations), the indication is that national emotions and involvement are more than casual. Public opinion is being galvanised for some kind of direct and 'necessary' involvement or action. A cartographic history of American involvement in Vietnam or Israeli concern with its neighbours illustrate this point as well. Naturally, in a society with a controlled press, such studies are more difficult, but not impossible. There are often military journals or other publications used to communicate national policy.

A second method for determining ethological zones is the use of questionnaires designed to elicit attitudes towards specific locations. Such a questionnaire should have its respondents rank each country or location according to their perception of its importance to national defence or security.[1] For example, the area is:

1. Essential.   We cannot survive without this area. War is justified to protect, take or hold the location.
2. Important.   We should try to control the area or feature, but if it means all out war or possible attack on our homeland, we should probably withdraw.
3. Important but not essential.   We should not commit troops or become physically involved in a potential conflict or war. But we should try to be of help in resolving the problem.
4. Not important.   We should not involve ourselves other than diplomatically or casually.

One need only review such contemporary situations as the U.S.S.R. in Poland and Afghanistan, the British and Argentinians in the Falklands or the Israeli behaviour towards Lebanon, Syria, Jordan and Egypt to see the implications and value of this model.

Table 22.1 illustrates some of the associations between behaviour and its manifestations.

PREDICTING THE INTENSITY OF CONFLICT

As already noted, <u>where</u> a conflict occurs is a major factor in determining the intensity of emotions associated with a resulting conflict. If we plot the range of ethologically- based emotions for conflict on a two-actor graph, it is possible to visualise the relationships (Figure 22.1). Again, it must be emphasised that we are dealing with perceptions, not with objective realities. It is the perceptions of each actor that are important, not values assigned by some outsider.

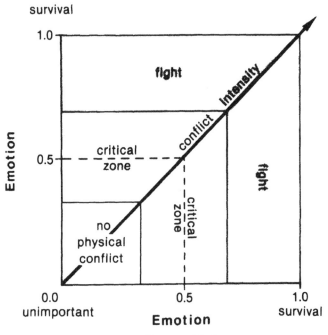

Figure 22.1

If we assume that each actor is of approximately equivalent strength (all things being equal, in other words), then we can see an interaction somewhat as each actor views the area as essential to its survival. The result is a bloody holocaust-type conflict in which one side <u>must</u> be defeated and probably destroyed. Certainly, the Roman-Carthaginian wars evolved to this stage, as have many recent conflicts.

If only one party views the area as essential to national survival and the other party views the location as only incidental to its national interest, then conflict is likely to be short-lived and the result to be in favour of the survivalist. This

TABLE 22.1

BEHAVIOUR MANIFESTATION	ZONE	RESULT
A cessation of press reports following a period of highly emotional threats.		
Elite military units or masses are moved to area.		
Detailed maps appear that stress the enemy's location and its threat to the state.	FIGHT	WAR
Nationalistic/patriotic themes dominate the press. Troops in battle dress seen with civilians and on city streets.		
Press reports refer to the 'enemy' in dehumanising terms		
Mobilisation of active units	CRITICAL - FIGHT	HIGH TENSION
Elite units placed in or near area		PROBABILITY OF WAR
Maps show general area of conflict		
Press comments stress seriousness of problem and dangers of war.		
Leaders issue warnings of dire consequences	CRITICAL -	NEGOTIATIONS
Troops told to be ready for possible mobilisation	FLIGHT	PROBABLE WITHDRAWAL
Only occasional references to area or conflict		
Officials and 'specialists' are interviewed for opinions.	FLIGHT	AVOIDANCE OF
Most people cannot locate the area on a map or explain problem.		INVOLVEMENT
Persons discussing possibility of conflict labelled 'warmongers'		(Unless 'ranking' motives triggered)

certainly was the case with the protracted conflict between the French and the Vietminh and between the U.S. and the Vietcong. It also explains the European and American 'actions' take in response to Russian presence in Poland, Czechoslovakia, Hungary and Afghanistan.

Naturally, should there be a great disparity in the technology or size of the opposing forces, this may turn the tide in favour of power versus emotion. However, as the conflict in Vietnam illustrated, superior technology alone is not sufficient when faced with a fight-to-the-death motivation.

When conflict occurs due to a 'ranking' motivation, emotions may appear as strong as when survival is at stake. The events of 1982 in the Falklands are illustrative. Still, if the area does not represent any survival threat, the conflict will remain low level and it is unlikely to evoke a do-or-die behaviour or intensity.

PROJECTING FUTURE POLICIES

If we apply this behavioural model in an attempt to predict future U.S. national policy, some of the more clear-cut fight zones would include America's immediate neighbours, Canada and Mexico, as well as the Caribbean and the Gulfs of Mexico and California. Naturally, the presence of any uninvited or hostile power in such areas would evoke immediate, drastic and militant responses with little hope of negotiation, as the Cuban missile crisis demonstrated. Less obvious would be the role of locations that provide raw materials to American industry and affect its economic stability. The most evident contemporary non- contiguous fight-zone for the U.S. is the Persian Gulf. Petroleum is popularly perceived as critical to the American way of life. Any threat to cut off or interrupt supplies evokes and has evoked an immediate and hostile response by the general public as well as the government.

However, possibly even more critical to U.S. survival than petroleum is a dependency on the strategic minerals listed below (Brown 1978).

Manganese	Cobalt	Titanium	Chromium
Aluminium	Tantalum	Platinum Group	Tin
Fluorspar	Nickel	Tungsten	Germanium-Indium
Beryllium	Zirconium	Industrial Diamonds	

The U.S. imports over 50 per cent of its needs of these minerals, over 75 per cent of the first eleven. Unlike petroleum which occurs naturally in the U.S., America must import many of these materials. Its industries cannot survive without them, even at a no-growth level.

What apparently is ignored - judging by actions and comments from high U.S. government officials - is that the vast majority of these minerals is concentrated in countries littoral to the South Atlantic. American relations with Brazil (manganese, tungsten, beryllium), Zaire (diamonds, tungsten, cobalt, manganese), Angola (diamonds, the platinum group) and especially South Africa (diamonds, tungsten, manganese, chromium, titanium) currently are critical to the survival of American industry. Thus, the South Atlantic forms a geographic area of maximum American political interest. Any threat by a foreign power to control or cut off these materials threatens the economic existence of the U.S. and thus must ultimately cause strong political or military reactions. There is, as yet, no emotional commitment, no sense of 'survival' associated with the South Atlantic.

We can also show how the perception of fight, critical and flight areas has been used to expedite political objectives in foreign negotiations. For example, the U.S. has applied significant, non-belligerent pressure on the Soviet Union for better accords by merely suggesting a U.S. sale of military technology to China. China is viewed as an area critical to Soviet Russia's survival. The possibility that China might become capable of a major war at a more rapid rate than is now occurring is a significant incentive to try and prevent this from happening.

CONCLUSIONS

Using a behavioural model of geopolitics, one can estimate the seriousness (and likely intensity) of various conflicts and treat them accordingly - rather than treating them as a major crisis.

When the U.S. and the U.S.S.R. confront each other in what appears to be a remote corner of the world, consider how serious the confrontation is likely to be in terms of the actual 'need' (versus prestige) of each country. Areas, such as Angola (industrial diamonds) that may seem distant and unimportant may, in fact, be critical zones. Wars that seem intense and emotional, such as Korea and Vietnam, may occur in flight zones and preparations for face-saving withdrawals should be considered, especially should the area be viewed as critical to another major power, such as China in the case of Korea. Awareness of the intensity of national behaviour and emotions associated with various geographic areas could help prevent blunders that often trigger accidental conflict.

Finally, when considering fight, critical and flight zones and their related behaviours, there is a need to be aware of the continual geographic shift of key areas. Today, the Mexican oil fields in conjunction with fields in Canada and Venezuela reduce the critical importance of the Middle East to U.S. survival. Breakthroughs in metallurgical technology could shift or even end dependence on various strategic metals. These in turn would result in redirected foreign policies.

end dependence on various strategic metals. These in turn would result in redirected foreign policies.

Regardless of such shifts in physical location, there always will be areas that are viewed as essential to national survival. If we are to control or minimise conflict, we must be aware of such geopolitical zones.

FOOTNOTE

1. Experience has demonstrated that the use of maps with the questionnaire is more confusing than helpful. Maps add variables (such as familiarity and ability to interpret scale and symbols) that tend to confuse.

REFERENCES

BROWN, G.S. [1977]
   United States Military Posture for FY 1978, (U.S. Government Printing Office, Washington, D.C.).
EIBL-EIBESFELDT, I. [1975]
   Ethology: The Biology of Behavior, (Holt, Rinehart and Winston, New York).
FOX, M.W. [1974]
   Concepts in Ethology: Animals and Human Behavior, (University of Minnesota Press, Minneapolis).
HALL, E.T. [1966]
   The Hidden Dimension, (Doubleday, New York).
HEDIGER, H. [1955]
   Studies of the Behavior of Captive Animals in Zoos and Circuses, (Criterion Books, New York).
HERZ, J.H. [1957]
   'Rise and demise of the territorial state', World Politics, 9, 473-93.
JONES, S.B. [1955]
   'Global strategic views', Geographical Review, 45, 492-508.
McCOLL, R.W. [1966]
   'Political geography as political ecology', Professional Geographer, 18, 143-45.

Chapter 23
AMERICAN FOREIGN POLICY FOR THE EIGHTIES

Saul B. Cohen

INTRODUCTION

While the founding fathers of modern Political Geography
are numbered among the creators of the field of International
Relations, their interest has been maintained only fitfully by
successor generations, especially those of the post-World War
II era. Karl Ritter, Friedrich Ratzel, Halford Mackinder and
Isaiah Bowman sought from their nineteenth and early twentieth
century vantage points to inject notes of realism into the conduct
of international affairs (Mackinder 1919).[1] They observed various
natural-conditioning factor concepts from which they developed
certain political world views and articulated consequent policies.
During the Inter-War periods, German Geopolitik emerged as a
distortion of the founders' theories. Karl Haushofer et al.
promulgated the pseudo-science that rationalised the conduct of
national policy as being naturally-determined, thus reversing
the logical relationship between cause and effect.

A World War Two generation of American geographers led by
Richard Hartshorne and Derwent Whittlesey repudiated the work
of these German geographers. In reaction to Nazi-serving
distortions, Hartshorne also influenced some American geographers
to shy away from the search for causality between the nature of
the earth as the habitat of man, and consequent political policy
lessons that might be drawn. It was also during this period
that most American - and indeed Western - political geographers
not only withdrew from the field of International Relations, but
left geopolitical analysis to political and other social scientists
- an ironic note in the light of geographers' early contributions
to their own field, as well as to International Relations.

In fact, there are useful links between the two approaches
in Political Geography to develop geopolitical hypotheses grounded
in the 'real world', and at the same time to utilise some of the
morphological and functional approaches emphasised by those who
shied away from making policy judgements (Cohen 1973). In this
context, we must continually keep in mind that the objective
world cannot be separated from man's psychological and cognitive

images of this world (Sprout and Sprout 1965). Not everyone draws the same application from a geopolitical theory, for each one of us applies unique value systems - national, cultural, ideological - to the theory. What is important, however, is the development of a natural conditions theory around which there can be general consensus. Russians, Chinese, Americans, English, Israelis and Arabs may well apply such theory to fulfil their own national interests. But if there is common acknowledgement of the existence of something like geopolitical reality, then at least there is a basis for negotiation and reconciliation of various interests, and of bridging the gap between 'objective ' and 'subjective' environments.

Political geographers have a responsibility to contribute to their respective national foreign policy debating grounds, rather than to avoid the field of international relations. It is in this spirit that certain aspects of American foreign policy will be discussed.

THE NEW INTERNATIONAL GEOPOLITICAL SYSTEM

As the global system matures, it has become increasingly specialised, integrated and hierarchical. In this connection the new global order is affected by the emergence of a substantial number of regional powers. Major powers have to create new sets of relationships, not only amongst themselves, but with these regional or second-order powers to create a new basis for global geopolitical equilibrium (Cohen 1982) (Figure 23.1).

Thus, American foreign policy is beginning to focus on the emergence of new regional forces that are giving a different direction and content to the international system. These regional forces play a crucial role in the search for a new global geopolitical equilibrium. Tables 23.1 and 23.2, and Figure 23.1 include twenty-seven states which have been identified by the author as being perceived or aspiring to regional power status (Cohen 1982). There will be reference to many of these states in the ensuing discussion. One illustration of the significance of regional powers in world affairs is that of the twenty-two members of the Cancun Conference of Industrial and Developing Nations; ten were in the regional power class, while six came from major power groupings.

It is clear that the United States can rarely hope to enter into exclusive bi-lateral relations with any single regional power. Rather, these relations must take into account the needs and postures of other major powers, especially America's allies of the EEC and Japan, but also China and the U.S.S.R. Moreover, relations between the United States and any single regional power are often balanced by its needs for links to another regional power in the same part of the world. A primary objective of United States foreign policy, therefore, is to find ways of

TABLE·23.1

Number of 1st and 2nd Order Powers by Geopolitical Region

GEOPOLITICAL REGION	1st ORDER POWERS	2nd ORDER POWER CONTENDERS	2nd ORDER POWER PROSPECTS
Anglo-America & the Caribbean	1	3	1
South America	-	2	-
Maritime Europe & the Maghreb	1	1	3
Offshore Asia	1	1	0
Heartland and Eastern Europe	1	2	1
East Asia	1	-	-
South Asia	-	1	1
Southeast Asia	-	1	1
The Middle East	-	1	5
SubSaharan Africa	-	1	2

TABLE 23.2

Regional Power Rankings

(A) TOP THIRD (42 to 33 points): (42) India, (41) Brazil, (40) Australia, Canada, Sweden, (35) Israel, (34) South Africa, (33) Yugoslavia.

(B) MIDDLE THIRD (31 to 29 points): (31) German Democratic Republic, Nigeria, Turkey, Venezuela, Vietnam, (30) Mexico, (29) Argentina, Egypt.

(C) LOWER THIRD (26 to 20 points): (26) Algeria, Indonesia, Spain, (24) Cuba, Iran, Morocco, Poland, (23) Pakistan, (22) Iraq, (21) Saudi Arabia, (20) Zaire.

To assess second-order power states, 27 states were ranked on an ordinal scale. The ranking was based on twelve items derived from eleven different criteria, each item being assigned a one through four weighting.

sharing responsibilities with other powers, both of the first-
and second-order, in the resolution of regional and global
conflict.

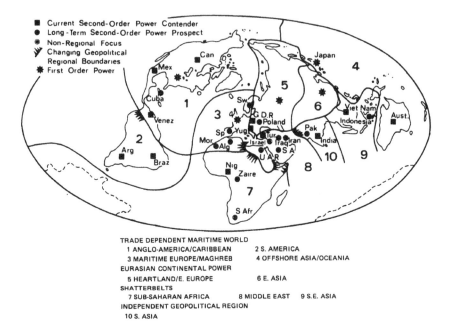

Figure 23.1: First and Second Order Powers.

   Shared responsibilities with major partners means that the
United States cannot go it alone in the Middle East. Long- term
access to Middle Eastern oil is more important to Common Market
Europe and Japan than it is to America. As distasteful and
perhaps illusory as certain European and Japanese initiatives
may seem relative to the Arab-Israeli conflict, or to events in
Iran, Afghanistan or Lebanon, these initiatives cannot be dismissed
out of hand. Increased European-Japanese policy initiatives in
the Middle East should be expected and, indeed, encouraged, for
the United States cannot speak of global partnership and at the
same time quash regional initiatives by its major allies.
   Moreover, just as it is in America's vital interest to make
greater common energy cause with Canada, Mexico and Venezuela
to lessen its dependence on Middle Eastern petroleum, the European
Communities can be expected to hedge their Middle East dependence
through expanding their energy resource relationships with the
Maghreb, West Africa and the U.S.S.R., and Japan with Southeast
and East Asia, and probably Siberia.

298

American foreign policy reassessment is based upon a view of the world which recognises the significant changes that have occurred as a result of redistribution of global economic power, the political emergence of the former colonial world, and changes in the global military balance. This reassessment did not begin with the administration of President Reagan. Rather, it began during the era of President Carter - with such actions as the Panama Canal divestiture, the emphasis on human rights, the ill-fated attempt to limit arms transfers, and the recognition of the importance of regional powers. While in some measure the Reagan administration has sought to roll back the clock and return America to the Cold War and Containment, in other and more basic ways, it has reinforced the new direction.

Surely the emphasis on defence spending that attempts to redress past weaknesses, the attempts to block the Soviet pipeline project, and the soft-pedalling of the human rights issue represent changes in substance and tone. But the threads of continuity between these two, and to a certain extent the Ford-Nixon administrations, are strong, and show consistent evolutionary change - e.g., the increased focus on the role of regional powers in addressing international issues; the search for military parity with the Soviet Union; the struggle to achieve a better partnership of interest with Maritime Europe; the new links to China; the concern with economic and political stability in Latin America; and the promotion of peace in the Middle East.

A nation's foreign policy is all-too-often judged for consistency by whether the policy is applied to _all_ other nations, or by whether the policy towards one particular state is constant over time. Thus, American foreign policy is deemed inconsistent if it applies a human rights standard to Chile and not to the Soviet Union, or if it promises Israel not to sell F-15 fuel-enhancement equipment to Saudi Arabia and then turns around within a matter of three years and agrees to do so.

However, there is another measure of foreign policy consistency that is even more critical. This has to do with relating domestic practices and principles to those that are pursued to foreign affairs. Think for a moment of the record of American inconsistency in this realm - the ringing proclamation by Franklin D. Roosevelt of the Four Freedoms for the world at a time when Jim Crowism was rampant and 10 per cent of the nation consisted of 'Invisible Men'; the commitment to a policy that opposes nuclear arms proliferation elsewhere in the world, with little regard for American nuclear arms self-restraint; the belief in free trade, but insistence that American industry and shipping be protected. With all of this, American leaders have become more self- conscious about failures to achieve consistency. Such foreign policy issues as free trade, human rights, the International Law of the Sea, and freedom of international movement have their roots in the free enterprise system and an open economy, and are, therefore, consistent with the nation's values and principles (Gottmann 1982).

The line between flexibility and inconsistency can, of course, be thin. In the best sense, flexibility represents a nation's abilities to adapt its ideological principles to fundamental changes in the environments within which it operates, and to change policies accordingly. As the stock of strategic raw materials within the United States has decreased, twin principles of conservation and inter-dependence have begun to replace age-old American principles of ever-increasing production, uninhibited consumerism, and self-sufficiency. Or, as the world has become far more complex than was its bi-polar character in the 1950s, the United States has backed off from the tenet that demanded of all states that they choose sides. India's neutrality and attempts to reassert Third World leadership are respected, and should China reestablish a modus vivendi with the Soviet Union, the United States surely would be satisfied with a Chinese posture of neutrality and open relations with the United States and the U.S.S.R., as well as with the Third World. America is seeking positive working relationships with Angola and Zimbabwe, hoping that their radical regimes will steer a neutral course in world affairs. Its search for similar solutions in Afghanistan and Kampuchea is another case-in-point. Nevertheless, the line between flexibility and inconsistency continues to be breached. Recent examples are the administration's volte-face on Nicaragua and its inability to accept a Nicaragua-type solution in El Salvador; its stance on the sale of AWACs to Saudi Arabia; and the relaxation of domestic energy conservation efforts.

There are a number of environments which influence the principles behind foreign policy. The United States operates within three enviornments: its internal or domestic scene; the various geopolitical regions into which the world is divided; and the global - or geostrategic scene. These are overlapping environments, of course, but differences of scale mean differences of texture and content. Also, the United States exercises varying degrees of control over these three worlds. Until the Second World War, America had been largely concerned with its domestic world. For the next three decades, focus was on the global. In the last few years, it has come to realise that the world is becoming hierarchically-organised, and that different regions are emerging as active participants in the global colloquy.

The fact that the United States, and indeed most nations, must deal with three environments, especially the regional, is cause for reconsideration of national principles in a very basic way, and for recasting of certain foreign policy postures. For example, America's sense of concern for events in the Caribbean, be this expressed through hostility towards Cuba or through efforts to assure Mexico's economic stability, in part because of the importance of the latter's petroleum resources and other strategic minerals, has influenced the recent liberal nature of U.S. immigration practices. This concern is at the root of the attempt to find a balance between those who wish to close the borders to immigration on the basis of high domestic unemployment, and those who would legalise the flow of immigrants who came more for economic than for political purposes. There is clearly great sensitivity to developing a new immigration policy that will close the flood gates without undermining America's political relations with its regional neighbours.

The moral commitments developed as a consequence of the war in Vietnam are another example of how external events profoundly affected United States immigration policy. Commitments to Vietnamese, Laotians and Cambodians for their roles in the war in Southeast Asia, as well as alliances with Taiwan and South Korea account for policies which have encouraged the immigration of Asians to the United States.

Another example of foreign policy tension is the American adherence to the principle of resolution of conflict through peaceful means. Peace-keeping efforts in Lebanon and the efforts to defuse conflict in Namibia are examples of this commitment. On the other hand, as alliances with regional powers become shored up - e.g., Pakistan, Egypt, Saudi Arabia, Israel, Morocco, South Korea, Turkey and Chile, the ever-increasing arms transfers are more likely to be used in intra-regional conflict than for the stated purposes of these transfers - the strengthening of these countries to ward off Soviet penetration into their regions.

ANGLO-AMERICA AND THE CARIBBEAN

Within this geopolitical region there are four second- order powers. The three that are allied with the United States have the capacity to provide it with nearly all of its energy import requirements. As it is, Mexico now ranks first and Canada fourth to fifth among U.S. oil suppliers. Venezuela could easily redirect its oil export flow to the North American market. Cuba, dependent upon the Soviet Union for massive economic and military aid, cannot be certain that it will not some day be left high-and-dry by a Soviet-American agreement to recognise United States primacy within its own geopolitical region in exchange for Soviet sway over certain bordering states like Afghanistan.

In recent years, the United States has begun to focus considerable efforts on the region within which it is situated, and to make significant policy readjustments towards those of its neighbours who are regional powers. In response both to Cuba's links to the U.S.S.R. and the world petroleum crisis, America has sought to forge a new pattern of relationships with Canada, Mexico and Venezuela. These patterns are based upon the concept of shared responsibility and on the search for consistency in foreign policy.

Clearly, these three regional powers have strong national interests that must be addressed. Venezuela and Mexico have over-extended their international indebtedness. They need to have their loans refinanced. They also must have stable and high commodity prices, access to technology and markets and, for Mexico and Canada, ease of movement of peoples across land borders. The American quid pro quo for support is stable energy supplies, security of capital investment, and political support to contain Soviet-Cuban influence within the region.

The success of negotiations between the United States and Canada over such issues as fishing rights, U.S.-generated trans-border environmental pollution, the role of 'Canadianisation' of American oil companies operating in Canada, and the pricing of Canadian natural gas exported to the United States is important, not simply to resolve individual problems, but to help shape a new partnership between the two countries. For America, Canada has become far more than the historic bridge to Great Britain, and an important go-between within the international community. Rather, Canada is its closest political and cultural friend, having, moreover, become a significant bridge to the English-speaking Caribbean nations.

In the context of giving the highest priority to strengthened political and economic ties with Caribbean second-order powers and Canada, the United States should seek ways of encouraging these allies to take the lead in regional conflict resolution. Venezuelan initiatives helped bring about a Panama Canal solution. So were its actions and those of Mexico useful in influencing America to accept change in Nicaragua.

Moreover, the Caribbean Basin Initiative by the United States, Canada, Mexico and Venezuela is a common approach to economic development. If successfully implemented, this programme could develop into a most striking model of how major and regional powers can work together in the years to come among the region's thirty-eight smaller and poorer countries. However, if the Reagan administration's emphasis upon trade is to be consistent with American values, it must deliver on its promise of one- way preferential duties and reduction of import quotas for the Caribbean nations. Moreover, aid to these governments to carry out large-scale transportation and power projects cannot be wished away. Giving grounds for optimism that the initiative will overcome temporary fiscal difficulties is that the three second-order powers are behaving as full partners - Venezuela and Mexico each have promised $350 million in annual aid to the region through a joint oil facility, matching America's yearly aid of $350 million, and Canada has doubled its promised aid figure to $100 million.

American policies towards Cuba can benefit from Mexican mediation. Thus, Mexican provision to Cuba of oil exploration technology can only result in lesser dependence of Cuba upon the Soviet Union. Further, Mexico's sympathy for the Nicaraguan and Salvadoran revolutionary movements can provide the latter with an option to dependence upon Cuba. The real question is whether America should be concerned about Nicaragua and El Salvador from the standpoint of its own security, or how events in these two countries might affect Mexico or Venezuela. If the answer is the latter, then the United States should welcome the assumption of Caribbean political initiatives by these two regional powers. It should take direct military action only when Mexico and Venezuela themselves are in jeopardy, or when a democracy like Costa Rica is economically or militarily endangered. Also, Mexico and Venezuela should be encouraged to take a hand in resolving the territorial conflict between Guatemala and Belize. That the two regional powers recognise the increasing burden of their responsibilities may be seen from their efforts to strengthen themselves militarily to take on regional defence roles, as well as from their regional economic efforts.

SOUTH AMERICA

Of the two second-order powers within this region, Brazil and Argentina, the former is the more important. There are many parallels between Brazil and Mexico, such as their size, resource base and regional weight. But there are differences, too. While Mexico has emerged from isolation to turn first to regional affairs, Brazil has pursued its global links and ambitions more assiduously than its regional aspirations. Brazil did not have to abandon a tradition of isolation and break the chains of twentieth century foreign domination, as did Mexico. Brazil's thirst for capital investment and technology, as well as its dreams for achieving world power status, have turned its attentions first and foremost to relations with global powers, especially the United States, Germany and Japan.

The cultural gap between Brazil and its neighbours, as well as the physical barriers of the Amazon Valley and the Andes, have so far limited Brazil's regional influence. However, the increasing lead that Brazil holds over Argentina strengthens the former's primacy in Eastern South America. Moreover, eventual development of Amazonas and the Guiana highland frontier will bring Brazil into closer contact with its Andean and Caribbean neighbours.

Brazil is America's logical ally in this region. Argentina, on the other hand, is far less suited to playing the role of regional power. Its military debacle in the Falklands, its consistent violation of human rights and its aggressive stance on its boundary quarrel with Chile all limit the role that Argentina can play in promoting Latin American stability. The ill-fated effort of Washington to use Argentina as a surrogate in Central America is an example of an ill-conceived and inconsistent American foreign policy. Surely a distant military dictatorship is no alternative to such regional allies in Central America as Venezuela and Mexico.

The United States has encountered difficulties in its relations with Brazil because of attempts to monopolise Brazil's external links, based upon the pattern of bi-lateral relations that emerged during World War II. However, America has not sufficiently appreciated the problems of Brazil's energy dependence. The turn by Brazil to West Germany for purchase of nuclear reactors, fuel processing plants and technology best serves Brazil's interests, given American policies on sale and control of nuclear technology. The fact that the United States has had to step in to help Brazil refinance its massive foreign debt, highlights American inconsistency with respect to actions that limit development of Brazil's nuclear power industry. Given its location and traditions, Brazil, is a very unlikely candidate to join the nuclear arms 'Cold Warrior' nations.

THE MIDDLE EAST

In the Middle East, there are no clear-cut regional powers, though Iran and Egypt have the potential, while such aspirants as Iraq and Saudi Arabia do not. However, Iran's size, oil resources and fundamentalist fervour barely paper over deep social and political fragmentation. Egypt, with its large army, Sunni traditions and foreign currency-earning capacities, has many economic and social weaknesses, and has yet to re- establish its pre-Camp David role of Arab world leadership. Still, peace with Israel has freed Egypt to focus on internal development and to apply its military leverage elsewhere in the Middle East. Israel, the region's most powerful military nation and most stable society, is not fully a regional power because it is culturally and religiously isolated within the region.

Having lost influence in Iran, the U.S. has chosen to build up Egypt's regional power, supporting it with military and economic aid on a par with that supplied to Israel. Moreover, the U.S. perceives its long-term ability to protect Middle Eastern oil as resting on the strength of its relationships within the Arab world, not on Israel's military power. Therefore, Washington has placed its bets on Egypt, which cannot and will not in the long run disengage from the Arab world.

Beyond this, full peace with Egypt provides Israel not only a historic opportunity to remove a major military threat, but the opportunity to involve Egypt in mediating the conflict with other states in the region.

Another aspect of future relationships lies in the realm of military power. The Israeli notion that it can become the American military surrogate in the Middle East is overly sanguine. Egypt is the logical surrogate and ally for the U.S. in any Persian Gulf conflict. It has the manpower and would not - as Israel obviously would - be an unwelcome, alien force to Arab states that conceivably might seek protection. If Israel continues to emphasise its importance to the U.S. essentially on the grounds of military prowess and as a Cold War ally, rather than on the basis of intrinsic spiritual and political qualities, it will find itself at a competitive disadvantage with Egypt. Granted that Egypt cannot become a genuine regional power without peace and military support from Israel - but Israel has the added disadvantage that it cannot play a positive regional role without reaching an accommodation with the Palestinian Arabs.

So, new patterns have begun to unfold. American foreign policy is to encourage Egyptian influence in the Persian Gulf through cooperation with Saudi Arabia. It increasingly relies on Egypt in addressing the Palestinian problem, by promoting cooperation between that country, Jordan and Saudi Arabia. Indeed, the U.S. will probably try to engage European nations in the process, rather than, as presently, to rebuff them, as evidenced by the composition of the multi-national force in Lebanon. Further, it can be anticipated that the U.S., along with European countries and perhaps even the Soviet Union, will create a situation in which both Arabs and Israelis are pressed to negotiate peace in accordance with the United Nations formula as expressed in Resolution 242.

SUBSAHARAN AFRICA

In SubSaharan Africa, Nigeria, while lacking Brazil's cohesiveness, has no clear rival for regional primacy. America's past hopes that Zaire could achieve national cohesiveness and develop as a regional leader have been dashed. There may, however, be a danger in the United State's assuming too great a political stake in Nigeria, even though its oil exports are presently important to the American economy, helping, along with exports from Mexico, Great Britain and Canada, to minimise American dependence upon Saudi Arabian petroleum.

Certainly the United States should not bear primary Western responsibility in Africa. Instead, the lead role should be shouldered by the nations of the European Communities, which have greater dependence upon African trade and which share closer geographical, political and cultural bonds. In this process, Nigeria, the preeminent regional power, plays a significant role with the first-order powers in seeking to find solutions to the Angolan and Namibian conflicts.

Recognition that a parallelism of forces operates within SubSaharan Africa and the Middle East can be useful in shaping American foreign policy. Nigeria and Egypt are second-order states with deep regional roots. South Africa and Israel, although by far the strongest military and most economically advanced powers, lack second-order status because of their regional isolation. The West Bank and Namibia are tinder boxes and threats to regional stability and peace. Lesser regional states like Zimbabwe and Zambia, and Jordan and Saudi Arabia can play important moderating roles in the resolution of these regional conflicts. Moreover, Angola and Syria can be isolated as Soviet satellites, or weaned away from the U.S.S.R. However, it is to Nigeria and Egypt, nations without an overwhelming ideological stake in the regional conflicts, that the major Western powers must turn, to share in the responsibility for finding equitable political solutions.

SOUTH ASIA

United States policy in South Asia remains ambiguous. On the one hand, recognition of India's strength as a regional power has moved America to wean India away from dependence upon the Soviet Union and towards a policy of neutrality. On the other hand, the United States continues to view Pakistan as a Middle Eastern, rather than South Asian state. This creates a major contradiction. American military support of Pakistan serves to continue to push India, with its South Asian geopolitical orientation, towards the U.S.S.R. It is doubtful, however, whether the risk of destabilisation of the situation in South Asia is offset by the gains that may be derived from Pakistan's influence as an American Middle Eastern partner.

For the United States to rationalise arms buildups in Pakistan in terms of the nation's possible role as a counter to Soviet policy in Afghanistan, or as a military protector of conservative Islamic states in the Persian Gulf, is questionable policy. Indeed, Pakistan's interests in Afghanistan are largely concerned with maintaining influence over Afghan opposition forces so as to diminish prospects for a Pathan 'Pushtunistan' that would further dismember Pakistan. Moreover, Pakistan, with its latent anti-Americanism and its political instability, is not a reliable ally, and certainly Islamic fundamentalism is not a stable basis for alliance between Middle Eastern nations and the West.

Just as Saudi Arabia is not capable of substituting for Iran as the 'policeman' of the Gulf, neither is Pakistan. It is hardly the 'Eastern anchor of a strategic consensus' to protect the Persian Gulf oil states from the Soviet Union. The United States continues to fall into the trap of being manipulated by Pakistan, but India is the essence of South Asia. India is not irrevocably committed to the Soviet Union and has the capacity and ideological will to develop the region as a truly independent geopolitical region. America's national interests are to further this development.

OFFSHORE ASIA

It took the debacle of the Vietnam War for the United States to recognise that the lead in upholding the Maritime World's interests in Southeast Asia should be taken by Japan and Australia from their Offshore Asian bases. Within the region itself, the likelihood is that Vietnam will consolidate its position as a second-order power by dominating the Northern peninsular half. Indonesia lacks the geopolitical maturity to play an equally decisive role in the southern part of the region, and the ASEAN Pact will have limited strength until Indonesia achieves firm second-order power status to provide leadership to the pact.

Therefore, Australia and Japan, from their Offshore Asia regional bases, should be encouraged to play the key role in stabilising the Southeast Asian context. Elsewhere in Asia, the United States must find better ways of strengthening Japan's partnership of efforts with America in South Korea. United State's policy of arming China as a counter to the U.S.S.R. may boomerang, because a militarily strengthened China is much more likely to be a serious threat to Japan than to the Soviet Union. Finally, the American backing of Pol Pot against the current Vietnam-supported Kampuchean government is the height of cynicism and a distorted sense of realpolitik.

CONCLUSION

Because of sweeping global economic and political changes, the United States no longer is the world's dominant power - nor for that matter do the American and Soviet superpowers dominate bi-polar worlds. Rather, they are the two strongest of five major political and economic power centres - the European Communities, China and Japan being the others.

All these major (first-order) powers cannot operate independently of one another over the long term. Moreover, they must cooperate with certain regional powers, for the international geopolitical system is exceedingly hierarchical and regionally framed, as well as multi-polar.

Those who reshape American global strategy have become increasingly aware that the North-South dialogue has been cast too simplistically. The South or the Developing World is not a unified entity. It consists of diverse clusters of nations in different regions, with varying potentials and at various stages of maturity. The interests of each of these clusters is increasingly regional, and nations within them are organising themselves hierarchically. In addition to the likelihood that the dialogue will become more regionally-framed, there is the reality that East-West relations within the industrial developed world will retain their geostrategic or global-spanning primacy.

To envisage a world of equilibrium, American foreign policy-makers should adopt a new mental map of the world. This is not a world divisible in two, along traditional seapower and landpower lines. The Soviet Union is now a major maritime power, and to continue to preach containment of the U.S.S.R. from 'reaching warm waters' is to lock the stable after the horse has bolted.

The fact that the United States cannot confine the outreach
of the Soviet Union does not mean that it should concede supremacy
on the open seas. On the contrary, erosion of its naval strength
was America's single greatest strategic error of the 1960s and
1970s. The Reagan administration's commitment to a naval buildup
is recognition that U.S. vital interests are tied to its abilities
to secure the links of the Trade-Dependent Maritime World. Just
as America must be prepared to meet the Soviet challenge on the
open seas, so must it have the naval capacity to carry the
challenge to the Soviet Union in its northern Atlantic and Pacific
waters.

On land, the United States is not in a position to seek
military parity with the U.S.S.R. operating from its Continental
Eurasian Realm. American policy accepts the reality that it is
strategically handicapped in the face of various Soviet probes
and pressures around its Eurasian periphery, and tailors its
political actions and military strategies to this reality.

The Soviet Union, on the other hand, needs to invest in
superior land force numbers to defend itself simultaneously from
thrusts in any possible direction - from Europe, the Middle East,
Central Asia, East Asia and even the Arctic. The United States
needs sufficient land power which in combination with air and
sea power assures its ability to act decisively in those
geopolitical regions which are part of the Maritime World, and
in SubSaharan Africa. Buildups of the NATO conventional forces
to a parity level with those of the Warsaw Pact, and of the Rapid
Deployment Force for limited actions outside of Europe are valid
objectives. However, pressuring the European NATO allies to
accept modernised tactical nuclear weapons against their will
would seem to have a greater cost than benefit, because it
suggests that America is prepared to abandon the use of strategic
nuclear weapons as a deterrent to Soviet attack on Western
Europe.

Elsewhere on the 'World Island', in Southeast Asia and the
Middle East, the United States strategy cannot match the Soviet
Union's or China's land forces. In such regions, American naval
and air strength bears the lion's share of the burden. The
Trident submarines and the Stealth bombers - not B-1s which will
be antedated almost before they are activated, warrant the highest
military development priority. Clearly, a direct land engagement
with the U.S.S.R. in Southeast Asia and the Middle East would
be linked to confrontation in Maritime Europe and/or Offshore
Asia. In Maritime Europe, where the strategic goals are global,
not regional, NATO cannot risk anything less than conventional
land power parity with the Soviet Union. However, the decision
to achieve such parity is a decision to be made by America's
European partners, not by the United States alone.

In shaping foreign policy for the Eighties, the United
States is faced with overriding challenges. The scale of world
arms transfers - 100 billion dollars, of which 40 per cent is
by the United States, is escalating rapidly and in seemingly

uncontrollable fashion. It is a measure of the distortion of global values that arms transfers are equal in value to food transfers. Need for an American agreement with the Soviet Union to limit nuclear weapons is uppermost in the minds of large segments of the world's population who dread a nuclear holocaust. However, a superpower agreement to limit conventional arms transfers, particularly to Third World countries, may be even more pressing for world peace than limitation of nuclear arms.

Developing countries have tripled their defence budgets in the past few years and their armed forces now represent 60 per cent of the world's 25 million military personnel. The economic strength of the Developing World is being sapped, feeding inflation and unemployment and increasing the social deficits of these nations. In many ways, the American drive to expand arms transfers relates as much to its energy problem as it does to military issues. In order to recycle petrodollars, the United States has engaged in a vast arms transfer programme. Continued American efforts to reduce oil consumption and to find alternative sources of energy will benefit the poorest nations of the Developing World, as well as the Western economy. With decreasing dependence upon oil imports, the chances for the United States to obtain an agreement with the Soviet Union on limiting arms transfers to the Developing World increase, precisely because the U.S.S.R. does not have the petrodollar recycling problem that the United States has.

However, an accommodation between the United States and the Soviet Union is a necessary but not sufficient condition for establishing global geopolitical equilibrium; American foriegn policy for the Eighties must aim at the increased involvement of the other first-order, and key regional powers as partners in the search for world stability. It is in this way that a new system of dynamic equilibrium can be forged - a system that is flexible, interactive and stable, and a system that is based upon shared responsibilities and shared power.

FOOTNOTE

1. Mackinder's work was an especially compelling world-view that had foreign policy implications.

REFERENCES

COHEN, S.B. [1973]
Geography and Politics in a World Divided, Revised edition, (Oxford University Press, New York).

COHEN, S.B. [1982]
   'The new map of global geopolitical equilibrium', <u>Political Geography Quarterly</u>, 1, 223-41.
GOTTMANN, J. [1982]
   'The basic problem of political geography: the organization of space and the search for stability', <u>Tijdschrift voor Economische en Sociale Geografie</u>', 73, 340-49.
MACKINDER, H. [1919]
   <u>Democratic Ideals and Reality</u> (Holt, New York).
SPROUT, H. and M. SPROUT [1965]
   <u>The Ecological Perspective on Human Affairs</u>, (Princeton University Press, Princeton).

Chapter 24
OVERVIEW

Stanley Waterman and Nurit Kliot

The revival of political geography over the past decade is now a well-documented phenomenon. It is no longer the 'moribund backwater' to which Berry (1969) referred nor is it the 'phoenix' or 'dead duck' to which Muir (1976) drew attention. On the contrary, the emergence of Political Geography Quarterly, the publication of Political Studies from Spatial Perspectives (Burnett and Taylor 1981), of textbooks by Muir (1975), Muir and Paddison (1980), Taylor and Johnston (1979) and others, of a large number and variety of articles in a wide spectrum of journals (Kliot 1982) and a tentative programme for political geography (Political Geography Quarterly) point to a real and living bird in a healthy environment.

The healthy pluralism within political geography is brought about by the keenness and seriousness of the debate over the topics that come under the aegis of political geography and what, if anything, should provide the guiding lights for workers in the discipline. An agreed content matter might indicate maturity but also a certain stagnation of ideas; certainly, a pluralism of ideas enlivens debate and draws in persons from a wide variety of backgrounds, even those who do not necessarily consider themselves to be primarily political geographers or their field of research to be within the perimeter of political geography. The essay by Peter Taylor, distinguishing 'political' from 'Political' geographers and the three responses to it testify to this; the chapters by Fred Boal and David Livingstone and by Arnon Soffer are good examples, too.

This volume is about pluralism and political geography and the subtitle of the book - people, territory and state - represents the fundamental triad of modern political geography (Johnston 1981). The spatial overlap of cultural groups and its political expression is what contributes the life-blood of the discipline.

As has been pointed out in the Introduction to this volume pluralism is not a new topic within the field of geography. Geography is, by its very nature, a plural discipline, interfacing with the social and natural sciences and the humanities. Thus, political geography, as a branch of geography is a natural location for the study of those aspects of plural societies which have both a geographic or spatial element and a political cause or effect.

The pluralist nature of political geography itself is developed in the paper by Peter Taylor. In discussing the role that theory should play in political geography, he illustrates three different approaches to the discipline - what Taylor terms the traditionalist, the welfare and the radical-Marxist approaches to political geography. Although he claims to espouse the latter of this triad, he is willing to accept a pluralistic framework which provides room for persons whose understanding and use of theory is somewhat different from his own. The responses from Saul Cohen, Ron Johnston and Ray Hudson would indicate that the threefold division of modern day political geographers is not agreed upon by the political geographers themselves.

CULTURAL, SOCIAL AND POLITICAL PLURALISM

A distinction is often made between cultural, social and political pluralism (Boal and Douglas 1982). Types of cultural pluralism, based on racial, religious, linguistic or other ethnic attributes lend further detail to cultural generalities. More often than not, there is considerable overlap and intercorrelation among the different categories. For example, most residents of Quebec are French-speaking and Roman Catholic, but not all of Canada's Francophones or Catholics are Quebecers, and Quebec regionalists/ nationalists do not have a monopoly on separatist movements in Canada, as David Knight points out in Chapter 11. This lack of clarity can lead to gross error as one attribute is used as a surrogate for another even though this may be a gross inaccuracy, or worse, may be used to explain a cultural or political process, as Neville Douglas indicates in Chapter 7.

When we refer to cultural pluralism, we are referring to a more benign outcome of pluralism which is not directly expressed in terms of authority or power. But, more often than not, cultural pluralism is translated into a social and then political variety.

Thus, in Knight's discussion of Canada, the ethnic- religious and linguistic differences resulted initially in the creation of separate provinces within a federal structure in British North America. This political expression of pluralism with its origins in cultural differences, in turn leads to the break-up of the state and to changes in the map of North America. Similarly, Arnon Soffer (Chapter 9) shows the existence in Northern Israel of two distinct cultural groups, Arabs and Jews, which, as a result of Israeli settlement policies now inhabit the territories which are located close to one another within a single state. Although they live and work together, regional circumstances

312

force each to look towards different core areas, the Jews to the Zionist Israeli state, the Arabs towards Palestinians living on the West Bank and to other Arabs in general. This example is elaborated upon by Gwyn Rowley in Chapter 15 who in discussing the territorial competition between these same two groups in the West Bank shows how the perception by certain Zionists of a place whose borders extend beyond those of the state has caused a radical change in the settlement geography of the region, and may in the future cause revolutionary changes within the Palestinian movement. The third example of pluralism in the Arab-Jewish context is provided by Moshe Brawer (Chapter 13).

Yet another example of the overlap between cultural and political pluralism has been provided by Fred Boal and David Livingstone in Chapter 12. The introduction of Protestant planters into seventeenth century Northern Ireland created a culturally plural society where a native Irish society - the most intensely Irish in Ireland - had previously existed. Over the years, the plural society was institutionalised, where the Protestant majority was elevated to the permanent situation of power-holders and the Catholic minority removed from any function in power-sharing and authority in the province. This situation, magnified in the microcosm of the Shankill-Falls Divide in Belfast, illustrates the feature referred to as 'multiple-minorities', in which each cultural group observes itself as a minority several times over, with concomitant difficulties.

BOUNDARIES: TOWARDS AN ANTITHESIS TO PLURALISM

The essence of plural societies, and plural states is a living together, a sharing of a common territory by groups which differ from one another in different ways. Thus, boundaries which separate groups and subdivide territories into discrete units, (like nationalism which creates nations from peoples), be they states, autonomous regions or adminstrative areas, represent the inability of the peoples in a given region to exist within a plural framework.

The presence or introduction of boundaries in any given area creates a dual situation. On the one hand the populations of the individual areas existing or created as a result of a boundary feel freer than those who must share their territory with others. On the other hand, freedom of movement of people, goods, and ideas is restricted or otherwise inhibited by the presence of this boundary. In many ways, a boundary might mark the failure of the pluralist ideal. It is commonly felt that the most marked effects of boundaries are at the international level. Moshe Brawer (Chapter 13) describes the different development of initially very similar populations on either side of a fully sealed border which existed for 18 years, and which changed its function to that of an administrative divide. As a result of this change of status, the population on both sides of the border began to move from their polar positions along the integration-division scale as Jews employed West Bank Arabs in industry within Israel only a few kilometres away. Similarly, Nurit Kliot's paper on losing a sense of place as a result of the signing of a peace treaty, in this case between Israel and Egypt, throws a slightly different light on this aspect of political geography.

Henk Reitsma's study on the development problems of land-locked countries in Africa puts pluralism in a special perspective. Looked at on a world scale, the continents and major regions are as actors on a plural world-stage. The states of Africa must share a continent. In this sense, Africa is a major region with a large number of nations and states sharing a single territory. As a result of tribal subdivisions, colonial legacies and a general unwillingness to change the status quo (as David Knight would have it, a rigid-state-structured world, (Chapter 11), the political borders divide Africa into approximately 50 sovereign units some of which have no access to the sea, and are thus dependent on the good will and acumen of their coastal neighbours for most of their import and export business.

The basic conservative nature of boundaries, once brought into existence is also noted in the section on changing jurisdiction. Rex Honey (Chapter 18) emphasises the basic dilemma of economic efficiency versus tradition, versatility versus continuity. People and institutions align themselves culturally, socially and economically with an existing set of boundaries until circumstances force them to consider the situation anew. As Knight (Chapter 11) writes in our age, the state makes the nation and by projection the existence of a clearly defined regional area helps create local loyalties. Attempts to alter the existing pattern can create political and social difficulties as Honey illustrates with examples from Wales and the United States. Andrew Burghardt (Chapter 17) graphically illustrates the difficulties encountered in Southern Ontario when economics and history clash, when planners and administrators were confronted by massive opposition from the public, especially in rural areas which would be hardest hit as a result of changes. At another scale Ronan Paddison (Chapter 19) shows the difficulties encountered by the Australian federal government under a Labor administration when it attempted to strengthen its position vis a vis the state governments by creating a third level between local and state government which would overrun existing state boundaries and which would be more directly dependent upon the federal government for financial support.

SCALES OF PLURALISM

As has already been noted, the definition of whether an area contains a plural or uniform society or whether a society is plural or uniform depends to a large degree on the scale of

observation. This is the essence of the studies by David Knight
in his discussion of nationalism versus regionalism and Arnon
Soffer in his study of majority/minority amongst Arabs in Galilee.
Thus, our attempt to view Jewish society in Israel depends on
the scale and generalisations in the study. Yosseph Shilhav
shows pluralism at the micro-scale, in this sense in the context
of Jerusalem. Ultra-Orthodox Jews use public space to stress
their cultural and housing problems. In order to ensure that
their conception of the world is not harmed and at the same time
working towards the creation of uniformly orthodox areas, the
ultra-Orthodox of Jerusalem have succeeded in segregating
themselves within a discreet ghetto area, while threatening
public space so as to ensure future additional space for themselves.
Likewise Boal and Livingstone bring to light the sharp divide
amongst Protestants and Catholics, in Belfast (Chapter 12).

Most of the contributors deal with pluralism at an intermediate
scale, that of the national level, but several examine international
scales. The case of African land-locked countries has already
been discussed. Stanley Brunn and Gerald Ingalls (Chapter 21)
investigate the degree to which states vote in a similar manner
on different issues at the United Nations, that body which gives
best expression to the pluralism of views in the world. They
succeed in identifying certain visible voting blocs over the
years through an examination of five specific votes, but their
major conclusion is that the situation is quite fluid, many
states changing their 'allegiances' on different votes.

Robert McColl and Saul Cohen both look at the world stage
from a different vantage point to Brunn and Ingalls. McColl
(Chapter 22) applies a model from ethology to explain the behaviour
of states on the international scene. The conclusions at which
he arrives bear some similarity to those of Saul Cohen (Chapter
23) whose work is based on an analysis of geopolitical forces
throughout the world - especially where American world-views are
concerned.

TANGIBLE PLURALISM AND HIDDEN PLURALISM

Most commonly, plural societies are presented as those in
which the population is polyglot and ethnically varied. But a
society can be racially, linguistically and religiously uniform
yet plural in nature as a result of class and status differences.
The papers by Shalom Reichman and Paul Claval deal with aspects
of equity and effectiveness. Claval in Chapter 6 stresses the
dichotomous nature of equality and freedom - where apparently
greater equality amongst members of a population in terms of
opportunities for all, and greater equity in terms of goods and
services provided does not necessarily bring about greater freedom
- paradoxically, almost, there appears to be a trade-off between
the two. Thus, in a given society in which intangible pluralism
has all but been eliminated, there is a loss of freedom.

Reichman's paper in Chapter 16 complements that of Claval. Whereas Claval stresses the human side of freedom and equality, Reichman puts the emphasis on government manipulation of space principally through the design and redesign of administrative boundaries. This effectively alters the accessibility of segments of the population to certain goods and services creating different levels within the state. This is a point taken up again by Richard Morrill in Chapter 8 on the emergence of pluralism as a reaction to decades during which the 'melting pot' approach marked American society. He shows how overemphasis on the expression of cultural pluralism and the uniqueness of various groups can lead to a decline in services and to inequities in society, the same point brought home by Claval, if from a different angle.

The debate around Peter Taylor's opening remarks on the role of theory in political geography develops the problem of the extent to which we either accept or reject the current western capitalist state and world order. The radical approach to the world's problems is motivated by a desire for social and economic justice, to see that the distribution of resources and wealth is fair. Saul Cohen presents the case for the traditional approach, and in doing so, notes that the factor of resource allocation has lost its position of paramount importance in many parts of the world to be replaced by cultural and social parameters.

SUMMARY

The papers presented in this book represent only a small sample of the topics currently under study in Political Geography. They cover only some of the topics that can be investigated under the heading of Pluralism. Perhaps the most glaring omission is any discussion of the geography of voting and electoral geography, both of which provide material for the analysis of the extent to which political pluralism exists within a society. It was not our objective here to be all-encompassing nor was it possible.

What we have tried to do is to show the importance of the pluralism theme in Political Geography and in doing so, we would hope that we have shown that some researchers have already gone some way towards meeting the research agendas outlined by the editors in the first issue of Political Geography Quarterly.

REFERENCES

BERRY, B.J.L. [1969]
 'Geographical reviews', Geographical Review, 59, 450.

316

BOAL, F.W. and J.N.H. DOUGLAS (eds.) [1982]
    Integration and Division: Geographical Perspectives on the
    Northern Ireland Problem, (Academic Press, London).
BURNETT, A.D. and P.J. TAYLOR (eds.) [1981]
    Political Studies from Spatial Perspectives, (Wiley,
    Chichester).
COX, K.R. [1979]
    Location and Public Problems, (Maaroufa, Chicago).
JOHNSTON, R.J. [1981]
    'British political geography since Mackinder: a critical
    review', in A.D. Burnett and P.J. Taylor (eds.), Political
    Studies, 11-31.
KLIOT, N. [1982]
    'Recent themes in Political Geography - a review',
    Tijdschrift voor Economische en Sociale Geografie, 73,
    270-79.
MUIR, R. [1976]
    'Political Geography - dead duck or phoenix?', Area, 8,
    195-200.
MUIR, R. [1981]
    Modern Political Geography, Second edition, (Macmillan,
    London).
MUIR, R. and R. PADDISON [1981]
    Politics, Geography and Behaviour, (Methuen, London).
POLITICAL GEOGRAPHY QUARTERLY [1982]
    'Editorial essay - Political Geography, research agendas
    for the future', Volume 1, 1-17.
TAYLOR, P.J. and R.J. JOHNSTON [1979]
    Geography of Elections, (Penguin, Harmondsworth).

INDEX

318

319